THE ESSENCE OF

MEASUREMENT

THE ESSENCE OF ENGINEERING SERIES

Published titles
The Essence of Solid-State Electronics
The Essence of Electric Power Systems
The Essence of Measurement
The Essence of Engineering Thermodynamics

Forthcoming titles
The Essence of Analog Electronics
The Essence of Circuit Analysis
The Essence of Optoelectronics
The Essence of Microprocessor Engineering
The Essence of Communications
The Essence of Power Electronics

THE ESSENCE OF

MEASUREMENT

Alan S. Morris

Prentice Hall
LONDON NEW YORK TORONTO SYDNEY TOKYO
SINGAPORE MADRID MEXICO CITY MUNICH

First published 1996 by
Prentice Hall Europe
Campus 400, Maylands Avenue
Hemel Hempstead
Hertfordshire HP2 7EZ
A division of
Simon & Schuster International Group

Typeset in 10/12pt Times
by MHL Typesetting Ltd, Coventry

Printed and bound in Great Britain by
T J Press (Padstow) Ltd

Library of Congress Cataloging-in-Publication Data

Morris, Alan S., 1948–
The essence of measurement/by Alan S. Morris.
p. cm. — (essence of engineering)
Includes bibliographical references and index.
ISBN 0-13-371675-9 (pbk.)
1. Mensuration. 2. Engineering instruments. I. Title.
II. Series.
T50.M62 1996
681′.2—dc20 95-43743
CIP

British Library Cataloguing in Publication Data

A catalogue record for this book is available from the British Library

ISBN 0-13-371675-9

1 2 3 4 5 00 99 98 97 96

To Jane, Nicola and Julia

Contents

Preface

Following the success of the author's now well-established earlier text entitled *Principles of Measurement and Instrumentation*, many lecturers who have used that as a course text have commented on the need for a text on measurement that is of similar quality but that presents the material in a more condensed form in a shorter text which students with limited funds can afford. *Essence of Measurement* has therefore been written to fulfil this need.

The aim of this new book is to present the subject of measurement, and its use within instrumentation systems, as an integrated and coherent subject. Measurement has been of great relevance to mankind ever since the earliest days of human civilization, when it was first used as a means of quantifying the exchange of goods in barter trade systems. Today, measurement systems, and the instruments and transducers used in them, are of immense importance in a wide variety of domestic and industrial activities. The growth in the number and sophistication of instruments used in industry has been particularly significant over the past two decades as automation schemes have been developed. A similar rapid expansion in their use has also been evident in military and medical applications over the same period.

The material presented and the style of its delivery are designed to ensure that the reader gains a thorough understanding of the theoretical considerations that govern the choice and usage of suitable instruments in particular measurement situations. In support of this, worked examples are included throughout the text and sample problems for the reader to try are presented at the ends of chapters.

After a general introduction to measurement principles in Chapter 1, the text continues in Chapter 2 with a discussion of the static and dynamic characteristics of instruments. Static characteristics determine the steady-state performance of an instrument, in terms of accuracy, etc., whilst dynamic characteristics describe an instrument's dynamic behaviour following a step change in the measured quantity. Consideration of both types of characteristics is very important when choosing a suitable instrument for a particular application.

The subject of measurement errors is introduced in Chapter 3, where the aim is to consider the separate types of random and systematic errors in a cohesive and comprehensive way. Mathematically sound formulae are presented for calculating the total measurement error in a system containing two or more separate measurement components, and ways of reducing errors are discussed. Following this, the subject of instrument calibration and mechanisms for ensuring traceability to national reference standards are discussed in Chapter 4.

Chapter 5 looks at the various ways available for interpreting, processing and enhancing the quality of measurements. Bridge circuits are a particularly important subject within this and these are covered separately in Chapter 6.

The theory of electrical meters and a general coverage of electrical parameter measurement are presented in Chapter 7, which is followed in Chapter 8 by a treatise on data recording and presentation, including a discussion on regression curve-fitting techniques. Finally, in Chapter 9, some instruments which are commonly used for measuring various physical quantities are reviewed.

Whilst the motivation for producing this book has been to provide a supporting text for introductory courses in measurement and instrumentation at a budget price, the depth of coverage in many areas extends beyond the minimum level required for introductory courses and enables the book to be also used in support of courses which are more advanced. However, the amount of mathematical notation within the text has been deliberately minimized as far as possible to make the book suitable as a course text at all levels of further and higher education.

Besides this role as a student course text, the book is also of relevance to the practising instrumentation technologist who wishes to upgrade his or her knowledge about the latest developments in the theory of measurement principles. In addition, it is anticipated that the review of common instruments in Chapter 9 will provide industrial readers with useful revision and reference material.

CHAPTER 1

Principles of measurement

1.1 Introduction

The past decade has seen a large and rapid growth in new industrial technology, encouraged by developments in electronics in general and computers in particular. This decade of rapid growth, often referred to as the 'electronics revolution', represents an improvement in production techniques of a similar magnitude to that brought about by the Industrial Revolution in the last century. At the spearhead of this thrust forward has been the digital computer, this being an essential component in both hard production automation schemes and the even more advanced flexible manufacturing systems that are beginning to emerge.

The massive growth in the application of computers to process control and monitoring tasks has spawned a parallel growth in the requirement for instruments to measure, record and control process variables. As modern production techniques dictate working to tighter and tighter accuracy limits, and as economic forces limiting production costs become more severe, so the requirement for instruments to be both accurate and cheap becomes ever harder to satisfy. This latter problem is at the focal point of the research and development efforts of all instrument manufacturers. In the past few years, the most cost-effective means of improving instrument accuracy has been found in many cases to be the inclusion of digital computing power within instruments themselves. These so-called intelligent instruments therefore feature prominently in current instrument manufacturers' catalogs.

1.2 Early measurement units

Intelligent instruments represent the latest stage in the present era of measurement technology. This era goes back to the start of the Industrial Revolution in the nineteenth century when measuring instruments first began to be developed to satisfy the needs of industrialized production techniques. However, the complete history of measurement techniques goes back much further, in fact by thousands of years to the very start of human civilization. As humans evolved from their ape-like ancestors, they ceased to rely on using caves for shelter and to hunt and forage whatever they could for food, and started instead to build their own shelters and produce food by planting seed, rearing animals and farming in an organized manner. Initially, civilized humans lived in family communities and one assumes that such groupings were able to live in reasonable harmony without great arguments about who was working hardest and who was consuming the most. With the natural diversity of human talents, however, particular family groups developed particular specializations. Some communities would excel at farming, perhaps because of the arable quality of the area of land they occupied, while other groups might be particularly proficient at building houses. This inevitably led to family communities producing an excess of some things but having a deficit in others. Trade of excess production between family communities therefore developed naturally, and followed a barter system whereby produce or work of one sort would be exchanged for produce or work of another.

Clearly, this required a system of measurement to quantify the amounts being exchanged and to establish clear rules about the relative values of different commodities. These early systems of measurement were based on whatever was available as a measuring unit. For purposes of measuring length, for example, the human torso was a convenient tool, and gave us units of the hand, the foot and the cubit (length of the forearm). Such measurement units allowed an approximate level of equivalence to be established about the relative value of quantities of different commodities. The length and breadth of constructional timber, for instance, could be measured in units of feet, and cloth could be similarly measured in units of feet-squared. This allowed a basis for determining the relative values of timber and cloth for barter trade purposes. However, such a system was clearly unfair when a person with large hands exchanged timber for cloth from a person with small hands (assuming that each used his or her own hands for measuring the commodity being exchanged).

1.3 Modern measurement units

Although generally adequate for barter trade systems, the measurement units just described were imprecise, varying as they did from one person to the next. As trade became more sophisticated in the late 1700s, there was an urgent need to establish units of measurement which were based on non-varying quantities, and the first to emerge was a unit of length (the metre) defined as 10^{-7} times the polar quadrant of the Earth. A platinum bar made to this length was established as a standard of length in the early part of the nineteenth century. This was superceded by a superior quality standard bar in 1889, manufactured from a platinum-iridium alloy. Since that time, technological research has enabled further improvements to be made in the standard used for defining length. Firstly, in 1960, a standard metre was redefined in terms of $1.650\,763\,73 \times 10^6$ wavelengths of the radiation from krypton-86 in vacuum. More recently, in 1983, the metre was redefined yet again as the length of path travelled by light in an interval of 1/299 792 458 seconds.

In a similar fashion, standard units for the measurement of other physical quantities have been defined and progressively improved over the years. The latest standards for defining the units used for measuring a range of physical variables are given in Table 1.1.

Table 1.1 *Definitions of standard units*

Physical quantity	*Standard unit*	*Definition*
Length	metre	The length of path travelled by light in an interval of 1/299 792 458 seconds
Mass	kilogram	The mass of a platinum--iridium cylinder kept in the International Bureau of Weights and Measures, Sèvres, Paris
Time	second	$9.192\,631\,770 \times 10^9$ cycles of radiation from vaporized caesium-133 (an accuracy of 1 in 10^{12} or 1 second in 36 000 years)
Temperature	kelvin	The temperature difference between absolute zero and the triple point of water is defined as 273.16 kelvin
Current	ampere	One ampere is the current flowing through two infinitely long parallel conductors of negligible cross-section placed 1 metre apart in a vacuum and producing a force of 2×10^{-7} newton per metre length of conductor
Luminous intensity	candela	One candela is the luminous intensity in a given direction from a source emitting monochromatic radiation at a frequency of 540 terahertz (Hz $\times 10^{12}$) and with a radiant density in that direction of 1.4641 mW/steradian. (1 steradian is the solid angle which, having its vertex at the centre of a sphere, cuts off an area of the sphere surface equal to that of a square with sides of length equal to the sphere radius)
Matter	mole	The number of atoms in a 0.012 kg mass of carbon-12

The early establishment of standards for the measurement of physical quantities proceeded in several countries at broadly parallel times, and in consequence, several sets of units emerged for measuring the same physical variable. For instance, length can be measured in yards or metres or several other units. Apart from the major units of length, subdivisions of standard units exist such as feet, inches, centimetres and millimetres, with a fixed relationship between each fundamental unit and its subdivisions.

Yards, feet and inches belong to the imperial system of units, which is characterized by having varying and cumbersome multiplication factors relating fundamental units to subdivisions such as 1760 (miles to yards), 3 (yards to feet) and 12 (feet to inches). This imperial system is the one that was used predominantly in the United Kingdom until a few years ago.

The metric system is an alternative set of units which includes for instance, the unit of the metre for measuring length and its centimetre and millimetre subdivisions. All multiples and subdivisions of basic metric units are related to the base by factors of ten, and such units are therefore much easier to use than imperial units. However, in the case of derived units such as velocity, the number of alternative ways in which these can be expressed in the metric system can lead to confusion.

Table 1.2 *Fundamental and derived SI units*

Quantity	*Standard unit*	*Symbol*	*Derivation formula*
(a) Fundamental units			
Length	metre	m	
Mass	kilogram	kg	
Time	second	s	
Electric current	ampere	A	
Temperature	kelvin	K	
Luminous intensity	candela	cd	
Matter	mole	mol	
(b) Supplementary fundamental units			
Plane angle	radian	rad	
Solid angle	steradian	sr	
(c) Derived units			
Area	square metre	m^2	
Volume	cubic metre	m^3	
Velocity	metre per second	m/s	
Acceleration	metre per second squared	m/s^2	
Angular velocity	radian per second	rad/s	
Angular acceleration	radian per second squared	rad/s^2	
Density	kilogram per cubic metre	kg/m^3	
Specific volume	cubic metre per kilogram	m^3/kg	
Mass flow rate	kilogram per second	kg/s	
Volume flow rate	cubic metre per second	m^3/s	
Force	newton	N	$kg \cdot m/s^2$
Pressure	newton per square metre	N/m^2	
Torque	newton--metre	N·m	
Momentum	kilogram--metre per second	kg·m/s	
Moment of inertia	kilogram--metre squared	$kg \cdot m^2$	
Kinematic viscosity	square metre per second	m^2/s	
Dynamic viscosity	newton--second per sq metre	$N \cdot s/m^2$	
Work, energy, heat	joule	J	N·m
Specific energy	joule per cubic metre	J/m^3	
Power	watt	W	J/s
Thermal conductivity	watt per metre--kelvin	W/m K	
Electric charge	coulomb	C	A·s
Voltage, emf, pot. diff.	volt	V	W/A
Electric field strength	volt per metre	V/m	
Electric resistance	ohm	Ω	V/A
Electric capacitance	farad	F	A·s/V
Electric inductance	henry	H	V·s/A
Electric conductance	siemen	S	A/V
Resistivity	ohm--metre	Ω·m	
Permittivity	farad per metre	F/m	
Permeability	henry per metre	H/m	
Current density	ampere per square metre	A/m^2	
Magnetic flux	weber	Wb	V·s
Magnetic flux density	tesla	T	Wb/m^2
Magnetic field strength	ampere per metre	A/m	
Frequency	hertz	Hz	s^{-1}
Luminous flux	lumen	lm	cd·sr
Luminance	candela per square metre	cd/m^2	
Illumination	lux	lx	lm/m^2
Molar volume	cubic metre per mole	m^3/mol	
Molarity	mole per kilogram	mol/kg	
Molar energy	joule per mole	J/mol	

As a result of this, an internationally agreed set of standard units (SI units or Système International d'Unités) has been defined, and strong efforts are being made to encourage the adoption of this system throughout the world. The full range of fundamental SI measuring units and the further set of units derived from them are given in Table 1.2. Two supplementary units for measuring angles are also included within this table. The International Committee for Weights and Measures has also permitted the continued usage of certain other non-SI, metric units which are in widespread use. A good example of this is the bar (1 bar = 10^5 N/m^2) for measuring fluid pressure. Only SI or permitted non-SI units are used in this book.

1.4 Applications of measurement

Today, the techniques of measurement are of immense importance in most facets of human civilization. Present-day applications of measuring instruments can be classified into three major areas. The first of these is their use in regulating trade, applying instruments which measure physical quantities such as length, volume and mass in terms of standard units.

The second application area of measuring instruments is in monitoring functions. These provide information which enables human beings to take some prescribed action accordingly. The gardener uses a thermometer to determine whether the heat should be turned on in the greenhouse or the windows opened if it is too hot. Regular study of a barometer allows us to decide whether we should take our umbrellas if we are planning to go out for a few hours. While there are thus many uses of instrumentation in our normal domestic lives, the majority of monitoring functions exist to provide the information necessary to allow a human to control some industrial operation or process. In a chemical process for instance, the progress of chemical reactions is indicated by the measurement of temperatures and pressures at various points, and such measurements allow the operator to make correct decisions regarding the electrical supply to heaters, cooling water flows, valve positions etc. One other important use of monitoring instruments is in calibrating the instruments used in the automatic process control systems described below.

Use as part of automatic control systems forms the third application area of measurement systems. Figure 1.1 shows a functional block diagram of a simple feedback control system which is designed to maintain some output variable y of a controlled process P at a reference value r. The value of the controlled variable y, as determined by a measuring instrument M, is compared with the reference value r, and the difference e is applied as an error signal to the correcting unit C. The correcting unit then modifies the process output such that the output variable is given by $y = r$. The characteristics of measuring instruments in such feedback control systems are of fundamental importance to the quality of control achieved. The accuracy and resolution with which an output variable of a process is controlled can never be better than the accuracy and resolution of the measuring instruments used. This is a very

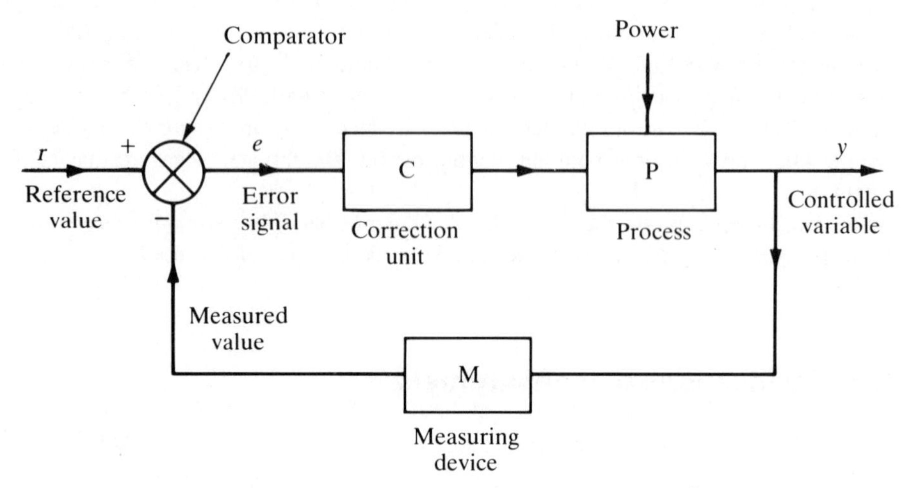

Figure 1.1 *Elements of a simple closed-loop control system*

important principle, but one which is often inadequately discussed in many texts on automatic control systems.

1.5 Measurement system components

A measuring instrument exists to provide information about the physical value of some variable being measured. In simple cases, an instrument consists of a single unit which gives an output reading or signal according to the magnitude of the unknown variable applied to it. However, in more complex measurement situations, a measuring instrument may consist of several separate elements as shown in Figure 1.2. These components might be contained within one or more boxes, and the boxes holding individual measurement elements might be either close together or physically separate. Because of this modular nature of the elements within it, it is common to refer to a measuring instrument as a measurement system, and this latter term is used extensively throughout this book to emphasize the modular nature.

Common to any measuring instrument is the *primary transducer*: this gives an output which is a function of the *measurand* (the input applied to it). For most, but not all, transducers, this function is at least approximately linear. Some examples of primary transducers are a liquid-in-glass thermometer, a thermocouple and a strain gauge. In the case of the mercury-in-glass thermometer, the output reading is given in terms of the level of the mercury, and so this particular primary transducer is also a complete measurement system in itself. In general, however, the primary transducer is only part of a measurement system. Some examples of primary transducers available for measuring various physical quantities are presented in Chapter 9. The output variable of a primary transducer is often in an inconvenient form and has to be

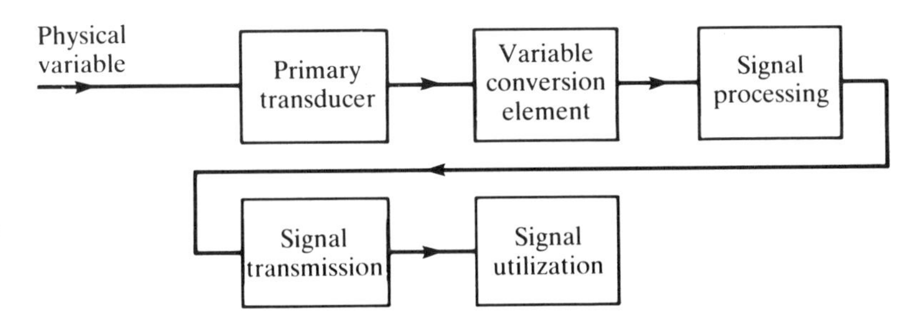

Figure 1.2 *Elements of a measuring instrument*

converted to a more convenient one. For instance, the displacement-measuring strain gauge has an output in the form of a varying resistance. This is converted to a change in voltage by a *bridge circuit*, which is a typical example of the variable conversion element shown in Figure 1.2.

Signal-processing elements exist to improve the quality of the output of a measurement system in some way. A very common type of signal processing element is the electronic amplifier, which amplifies the output of the primary transducer or variable conversion element, thus improving the sensitivity and resolution of measurement. This element of a measuring system is particularly important where the primary transducer has a relatively low magnitude output. For example, thermocouples have a typical output of only a few millivolts. Other types of signal-processing element are those that filter out induced noise and remove mean levels, etc.

The observation or application point of the output of a measurement system is often some physical distance away from the site of the primary transducer which is measuring a physical quantity, and some mechanism of transmitting the measured signal between these points is necessary. Sometimes, this separation is made solely for purposes of convenience, but more often it follows from the physical inaccessibility or environmental unsuitability for mounting the signal presentation/recording unit at the site of the primary transducer. The signal transmission element has traditionally consisted of single or multicored cable, which is often twisted and screened to minimize signal corruption by induced electrical noise. Now, optical fibre cables are being used in ever-increasing numbers in modern installations, in part because of their low transmission loss and imperviousness to the effects of electrical and magnetic fields.

The final element in a measurement system is the point where the measured signal is utilized. In some cases, this element is omitted altogether because the measurement is used as part of an automatic control scheme, and the transmitted signal is fed directly into the control system. In other cases, this element in the measurement system takes the form either of a signal presentation unit or of a signal-recording unit. These take many forms according to the requirements of the particular measurement application, and the range of possible units is discussed more fully in Chapter 8.

1.6 Instrument classification

Instruments can be subdivided into separate classes according to several criteria. These subclassifications are useful in broadly establishing several attributes of particular instruments such as accuracy, sensitivity, resolution[1] and cost, and suitability in different applications.

1.6.1 Active/passive instruments

Instruments may be classified as either active or passive according to whether the instrument output is produced entirely by the quantity being measured or whether the quantity being measured simply modulates the magnitude of some external power source. This is illustrated by the following examples.

The pressure measuring device shown in Figure 1.3 is an example of a passive instrument. The pressure of the fluid is translated into a movement of a pointer against a scale. The energy expended in moving the pointer is derived entirely from the change in pressure measured: there are no other energy inputs to the system.

The float-type petrol-tank level indicator, as sketched in Figure 1.4, is an example of an active instrument. Here, the change in petrol level moves a potentiometer arm, and the output signal consists of a proportion of the external voltage source applied across the two ends of the potentiometer. The energy in the output signal comes from the external power source: the primary transducer float system is merely modulating the value of the voltage from this external power source. In active instruments, the external power source is usually in electrical form, but in some cases it can be other forms of energy such as pneumatic or hydraulic.

One very important difference between active and passive instruments is the level of measurement resolution which can be obtained. With the simple pressure gauge shown, the amount of movement made by the pointer for a particular pressure change is closely defined by the nature of the instrument. While it is possible to increase measurement resolution by making the pointer longer, such that the pointer tip moves through a longer arc, the scope for such improvement is clearly restricted by the practical limit of how long the pointer can conveniently be. In an active instrument, however, adjustment of the magnitude of the external energy input allows much greater control over measurement resolution. While the scope for improving measurement resolution is much greater incidentally, it is not infinite because of limitations placed on the magnitude of the external energy input, in consideration of heating effects and for safety reasons.

In terms of cost, passive instruments are normally of a more simple construction than active ones and are thus cheaper to manufacture. Therefore,the choice between active and passive instruments for a particular application involves carefully balancing the measurement resolution requirements against cost.

[1] The terms accuracy, sensitivity and resolution are formally defined in Chapter 2.

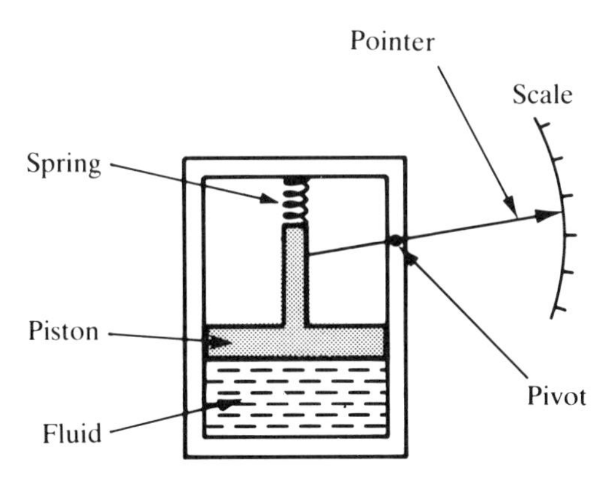

Figure 1.3 *Passive pressure gauge*

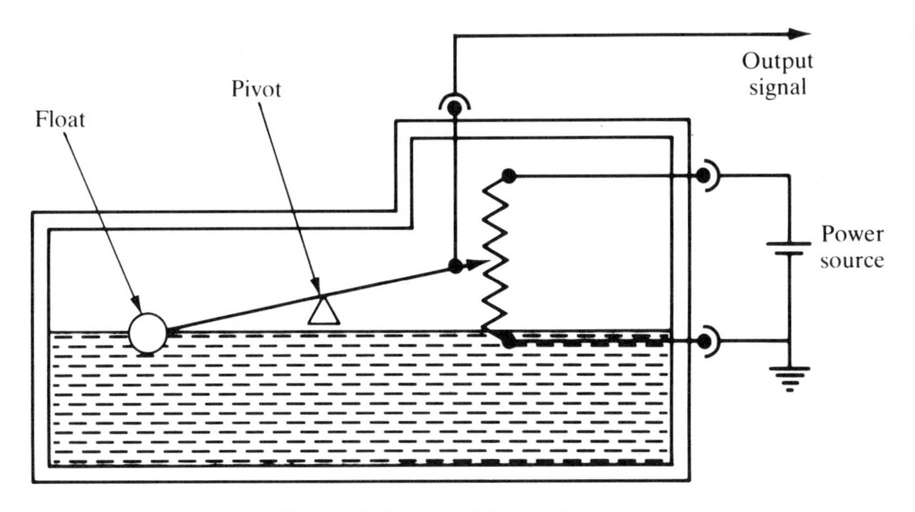

Figure 1.4 *Petrol-tank level indicator*

1.6.2 **Null/deflection-type instruments**

The pressure gauge just mentioned (Figure 1.3) is a good example of a deflection type of instrument, where the value of the quantity being measured is displayed in terms of the amount of movement of a pointer.

An alternative type of pressure gauge is the dead-weight gauge shown in Figure 1.5 which is a null-type instrument. Here, weights are put on top of the piston until the downward force balances the fluid pressure. Weights are added until the piston reaches a datum level, known as the null point. Pressure measurement is made in terms of the value of the weights needed to reach this null position.

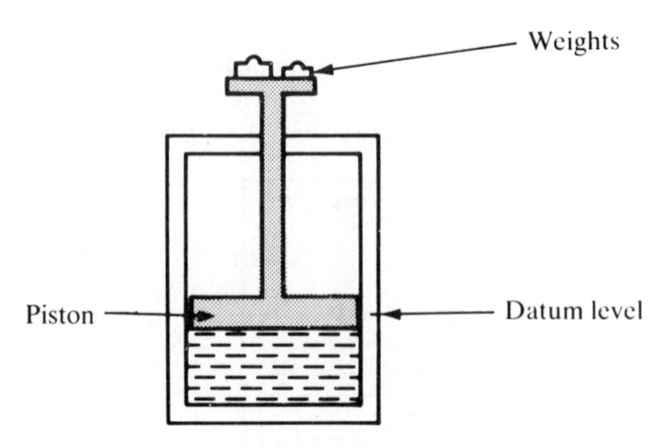

Figure 1.5 *Dead-weight pressure gauge*

The accuracy of these two instruments depends on different factors. For the first one, it depends on the linearity and calibration of the spring, while for the second, it relies on the calibration of the weights. As calibration of weights is much easier than careful choice and calibration of a linear-characteristic spring, this means that the second type of instrument will normally be the more accurate. This is in accordance with the general rule that null-type instruments are more accurate than deflection types.

In terms of usage, the deflection-type instrument is clearly more convenient. It is far simpler to read the position of a pointer against a scale than to add and subtract weights until a null point is reached. A deflection-type instrument is therefore the one that would normally be used in the workplace. However, for calibration duties, the null-type instrument is preferable because of its superior accuracy. The extra effort required to use such an instrument is perfectly acceptable in this case because of the infrequent nature of calibration operations.

1.6.3 ***Monitoring/control instruments***

An important distinction between different instruments is whether they are suitable only for monitoring functions or whether their output is in a form that can be directly included as part of an automatic control system. Instruments that give only an audio or visual indication of the magnitude of the physical quantity measured, such as a liquid-in-glass thermometer, are only suitable for monitoring purposes. This class normally includes all null-type instruments and most passive transducers.

For an instrument to be suitable for inclusion in an automatic control system, its output must be in a suitable form for direct input into the controller. Usually, this means that an instrument with an electrical output is required, although other forms of output such as optical or pneumatic signals are used in some systems.

1.6.4 *Analog/digital instruments*

An analog instrument gives an output which varies continuously as the quantity being measured changes. The output can have an infinite number of values within the range that the instrument is designed to measure. The deflection-type of pressure gauge described earlier in this chapter (Figure 1.3) is a good example of an analog instrument. As the input value changes, the pointer moves with a smooth continuous motion. While the pointer can therefore be in an infinite number of positions within its range of movement, the number of different positions that the eye can discriminate between is strictly limited, this discrimination being dependent upon how large the scale is and how finely it is divided.

A digital instrument has an output which varies in discrete steps and so can only have a finite number of values. The rev-counter sketched in Figure 1.6 is an example of a digital instrument. A cam is attached to the revolving body whose motion is being measured, and on each revolution the cam opens and closes a switch. The switching operations are counted by an electronic counter. This system can only count whole revolutions and cannot discriminate any motion which is less than a full revolution.

The distinction between analog and digital instruments has become particularly important with the rapid growth in the application of microcomputers to automatic control systems. Any digital computer system, of which the microcomputer is but one example, performs its computations in digital form. An instrument whose output is in digital form is therefore particularly advantageous in such applications, as it can be interfaced directly to the control computer. Analog instruments must be interfaced to the microcomputer by an analog-to-digital (A/D) converter, which converts the analog output signal from the instrument into an equivalent digital quantity which can then be applied to the computer input. This conversion has several disadvantages. Firstly, the A/D converter adds a significant cost to the system. Secondly, a finite time is involved in the process of converting an analog signal to a digital quantity, and this time can be critical in the control of fast processes where the accuracy of control depends on the speed of the controlling computer. Degrading the speed of operation of the control

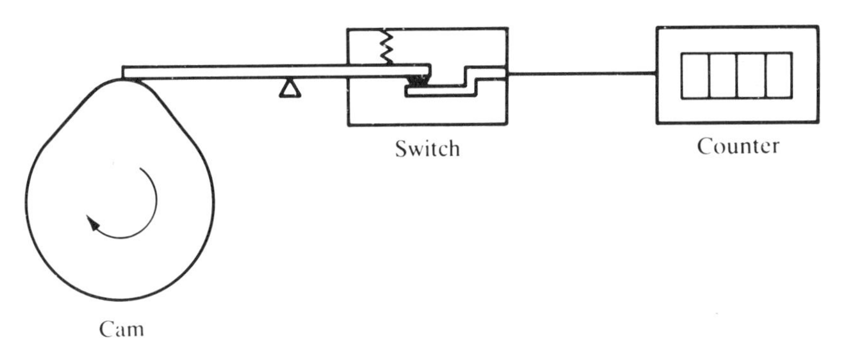

Figure 1.6 *Rev-counter*

computer by imposing a requirement for A/D conversion thus impairs the accuracy by which the process is controlled.

1.7 Measurement errors

Before closing this introductory chapter, mention must also be made of measurement errors. Measurement system errors can be divided into two categories, systematic and random errors. Various mechanisms exist for reducing these two types of error.

A characteristic feature of all systematic errors is that they produce errors which are consistently on the same side of the true value, i.e. either all the errors are positive or they are all negative. Systematic errors arise from many causes, which are discussed fully in Chapter 3. These include system disturbance due to measurement, environmental changes (modifying inputs) and drift in instrument characteristics. Large errors due to instrument characteristic drift are avoided by recalibrating instruments at suitable intervals (see Chapter 4). In the case of other sources of systematic error, a good measurement technician can largely eliminate errors by calculating their effect and correcting the measurements. This is done automatically by intelligent instruments.

Random errors are in many ways easier to deal with because they consist generally of small perturbations of the measurement either side of the correct value, i.e. positive errors and negative errors occur in approximately equal numbers for a series of measurements made of the same quantity. Therefore, random errors can be largely eliminated by averaging a few measurements of the same quantity. Unfortunately, averaging a number of measurements cannot be guaranteed to produce a value close to the true one because random errors occasionally cause large perturbations from the true value. Thus, it is necessary to describe measurements subject to random errors in probabalistic terms, typically that there is a 95% probability that the measurement error is within boundaries of ±1% from the true value. This is discussed more fully in Chapter 3.

References and further reading

ISO 1000, 1992 (BS 5555, 1993): Specification for SI units and recommendations for use of their multiples and of certain other units.

CHAPTER 2

Instrument characteristics

2.1 Introduction

The accuracy and performance of measurement systems are strongly governed by the characteristics of the instruments and transducers used within them. Knowledge of these characteristics is essential when designing measurement systems to ensure that the measurement requirements are met and that the most appropriate instruments or transducers are used in relation to the anticipated operating conditions of the system. The characteristics of any particular instrument or transducer are normally given in the data sheet for the instrument supplied by its manufacturer. It is important to note that the values quoted for instrument characteristics in such a data sheet only apply when the instrument is used under specified standard calibration conditions. Due allowance must be made for variations in the characteristics when the intrument is used in other conditions.

Instrument characteristics can be divided into two categories, static and dynamic. Static characteristics are those that describe an instrument's parameters (e.g. resolution and accuracy) in steady-state, i.e. when the instrument output has settled to a steady reading. The static characteristics of an instrument have a fundamental effect on the quality of measurements obtained from it. On the other hand, dynamic characteristics describe the dynamic response of an instrument between the time that a measured quantity changes and the time when the instrument output attains a constant value. The main consequence of dynamic characteristics is that, depending on the nature of the characteristics, a finite time must elapse between a measured quantity changing value and the instrument output being read.

2.2 Static characteristics

The various static characteristics of instruments are defined in the following paragraphs. More formal definitions can be found in the documents referenced (BS 5233, BS 5532 and ISO 3534).

2.2.1 Accuracy, inaccuracy and uncertainty

The term *accuracy* quantifies the degree of correctness of a measurement. A measurement with high accuracy will have very little error, whereas a measurement with low accuracy is likely to have a large error. Commonly, the word accuracy is used to quantify the maximum error which may exist in a measurement, though, strictly speaking, this quantifies the *inaccuracy* rather than the accuracy. The phrase *measurement uncertainty* is sometimes used instead of inaccuracy and means exactly the same. The misuse of the word accuracy to describe something which is really inaccuracy is widely met in instrument data sheets. In data sheets, accuracy is frequently quoted as a percentage of the full-scale reading of an instrument, but, if the figure is quoted as a few percent, then this most certainly refers to the inaccuracy rather than the accuracy. If, for example, a pressure gauge of range 0–10 bar has a quoted accuracy of ±1.0% f.s. (±1% of full-scale reading), then the maximum error to be expected in any reading is 0.1 bar.

Quoting accuracies in this way has important consequences when measurements are made which are small in magnitude compared with the instrument's range. When the pressure gauge just mentioned is reading 1.0 bar, the possible error is 10% of this value. For this reason, it is an important measurement system design rule that instruments are chosen such that their range is appropriate to the spread of values being measured, in order that the best possible accuracy be maintained in instrument readings. Thus, if we were measuring pressures with expected values between 0 and 1 bar, we would not use an instrument with a range of 0–10 bar.

2.2.2 Precision/repeatability/reproducibility

Precision is a term that describes an instrument's degree of freedom from random variations in its output when measuring a constant quantity. If a large number of readings are taken of the same quantity by a high-precision instrument, then the spread of readings will be very small.

Precision is often, though incorrectly, confused with accuracy. High-precision does not imply anything about measurement accuracy. A high-precision instrument may have a low accuracy. Low-accuracy measurements from a high-precision instrument are normally caused by a bias in the measurements, which is removable by recalibration.

The terms repeatability and reproducibility mean approximately the same but are applied in different contexts as given below. *Repeatability* describes the closeness of output readings when the same input is applied repetitively over a short period of time, with the same measurement conditions, same instrument and observer, same location and same conditions of use maintained throughout. Reproducibility describes the closeness of output readings for the same input when there are changes in the method of measurement, observer, measuring instrument, location, conditions of use and time of measurement. Both terms thus describe the spread of output readings for the same

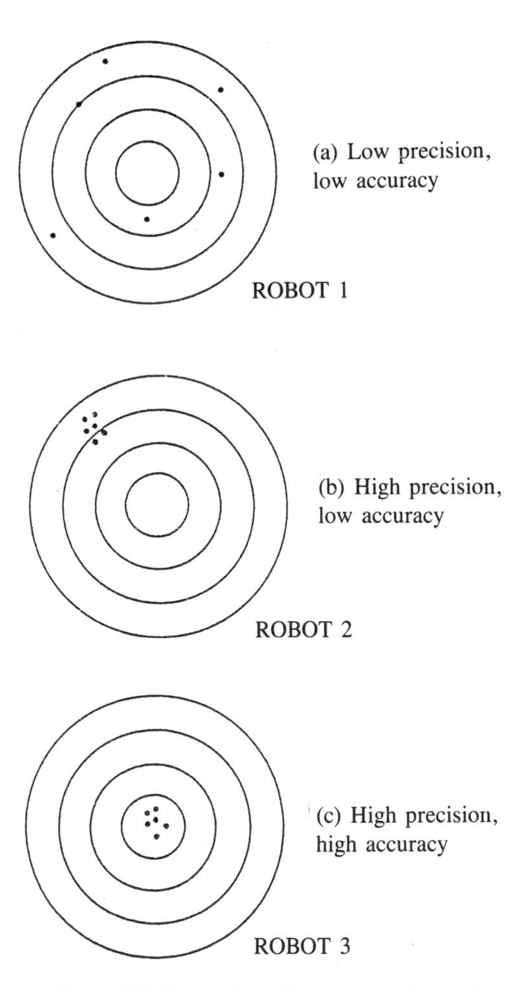

Figure 2.1 *Comparison of accuracy and precision*

input. This spread is referred to as repeatability if the measurement conditions are constant and as reproducibility if the measurement conditions vary.

The degree of repeatability or reproducibility in measurements from an instrument is an alternative way of expressing its precision. Figure 2.1 illustrates this more clearly. The figure shows the results of tests on three industrial robots which were programmed to place components at a particular point on a table. The target point was at the centre of the concentric circles shown, and the black dots represent the points where each robot actually deposited components at each attempt. Both the accuracy and precision of Robot 1 is shown to be low in this trial. Robot 2 consistently puts the component down at approximately the same place but this is the wrong point. Therefore, it has high precision but low accuracy. Finally, Robot 3 has both high precision and high accuracy, because it consistently places the component at the correct target position.

2.2.3 **Tolerance**

Tolerance is a term which is closely related to accuracy and defines the maximum error which is to be expected in some value. While it is not, strictly speaking, a static characteristic of measuring instruments, it is mentioned here because the accuracy of some instruments is sometimes quoted as a tolerance figure.

Tolerance, when used correctly, describes the maximum deviation of a manufactured component from some specified value. Crankshafts, for instance, are machined with a diameter tolerance quoted as so many micrometres (or microns, 10^{-6} m), and electric circuit components such as resistors have tolerances of perhaps ±5%. One resistor chosen at random from a batch having a nominal value 1000 Ω and tolerance ±5% might have an actual value anywhere between 950 Ω and 1050 Ω.

2.2.4 **Range or span**

The range or span of an instrument defines the minimum and maximum values of a quantity that the instrument is designed to measure.

2.2.5 **Bias**

Bias describes a constant error which exists over the full range of measurement of an instrument. This error is normally removable by calibration.

Bathroom scales are a common example of instruments which are prone to bias. It is quite usual to find that there is a reading of perhaps 1 kg with no one stood on the scales. If someone of known weight 70 kg were to get on the scales, the reading would be 71 kg, and if someone of known weight 100 kg were to get on the scales, the reading would be 101 kg. This constant bias of 1 kg can be removed by calibration: in the case of bathroom scales this normally means turning a thumbwheel with the scales unloaded until the reading is zero.

2.2.6 **Linearity**

It is normally desirable that the output reading of an instrument is linearly proportional to the quantity being measured. Figure 2.2 shows a plot of the typical output readings of an instrument when a sequence of input quantities are applied to it. Normal procedure is to draw a good fit straight line through the Xs. (While this can often be done with reasonable accuracy by eye, it is always preferable to apply a mathematical least-squares line-fitting technique, as described in Chapter 8). The non-linearity is then defined as the maximum deviation of any of the output readings marked X from this straight line. Non-linearity is usually expressed as a percentage of full-scale reading.

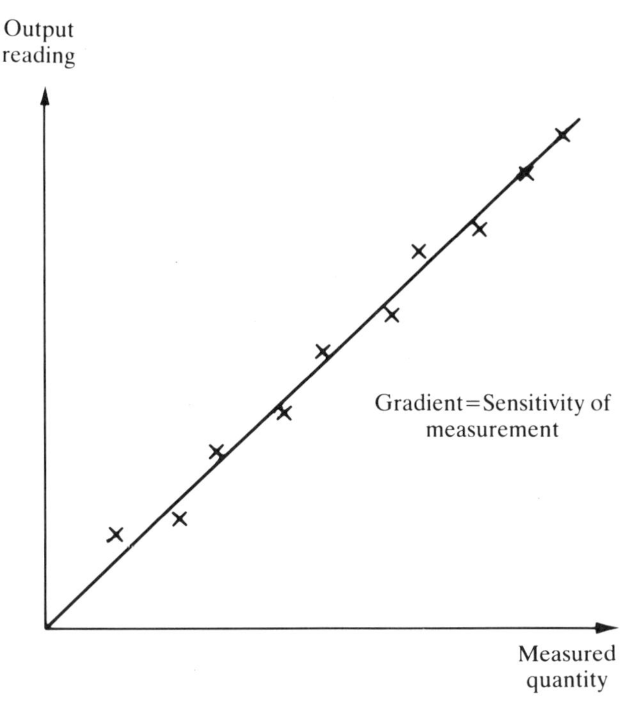

Figure 2.2 *Instrument output characteristic*

2.2.7 Sensitivity of measurement

The sensitivity of measurement is a measure of the change in instrument output which occurs when the quantity being measured changes by a given amount. Sensitivity is thus the ratio:

$$\frac{\text{scale deflection}}{\text{value of measurand causing deflection}}$$

The sensitivity of measurement is therefore the slope of the straight line drawn on Figure 2.2. If, for example, a pressure of 2 bar produces a deflection of 10 degrees in a pressure transducer, the sensitivity of the instrument is 5 degrees/bar (assuming that the deflection is zero with zero pressure applied).

EXAMPLE 2.1

The following resistance values of a platinum resistance thermometer were measured at a range of temperatures. Determine the measurement sensitivity of the instrument in ohms/°C.

Resistance (Ω)	Temperature (°C)
307	200
314	230
321	260
328	290

SOLUTION

If these values are plotted on a graph, the straight line relationship between resistance change and temperature change is obvious.

For a change in temperature of 30 °C, the change in resistance is 7 Ω. Hence the measurement sensitivity = 7/30 = 0.233 Ω/°C.

2.2.8 Sensitivity to disturbance

The calibrations and specifications of an instrument are only valid under controlled conditions of temperature, pressure, etc. These standard ambient conditions are usually defined in the instrument's specification. As variations occur in the ambient temperature, etc., certain static instrument characteristics change, and the sensitivity to disturbance is a measure of the magnitude of this change. Such environmental changes affect instruments in two main ways, known as zero drift and sensitivity drift.

Zero drift describes the effect where the zero reading of an instrument is modified by a change in ambient conditions. Typical units by which zero drift is measured are volts/°C, in the case of a voltmeter affected by ambient temperature changes. This is often called the zero drift coefficient related to temperature changes. If the characteristic of an instrument is sensitive to several environmental parameters, then it will have several zero drift coefficients, one for each environmental parameter. The effect of zero drift is to impose a bias in the instrument output readings: this is normally removable by recalibration in the usual way. A typical change in the output characteristic of a pressure gauge subject to zero drift is shown in Figure 2.3(a).

Sensitivity drift (also known as *scale factor drift*) defines the amount by which an instrument's sensitivity of measurement varies as ambient conditions change. It is quantified by sensitivity drift coefficients which define how much drift there is for a unit change in each environmental parameter that the instrument characteristics are sensitive to.

Many components within an instrument are affected by environmental fluctuations, such as temperature changes: for instance, the modulus of elasticity of a spring is temperature-dependent. Figure 2.3(b) shows what effect sensitivity drift can have on the output characteristic of an instrument. Sensitivity drift is measured in units of the form (angular degree/bar)/°C

If an instrument suffers both zero drift and sensitivity drift at the same time, then the typical modification of the output characteristic is as shown in Figure 2.3(c).

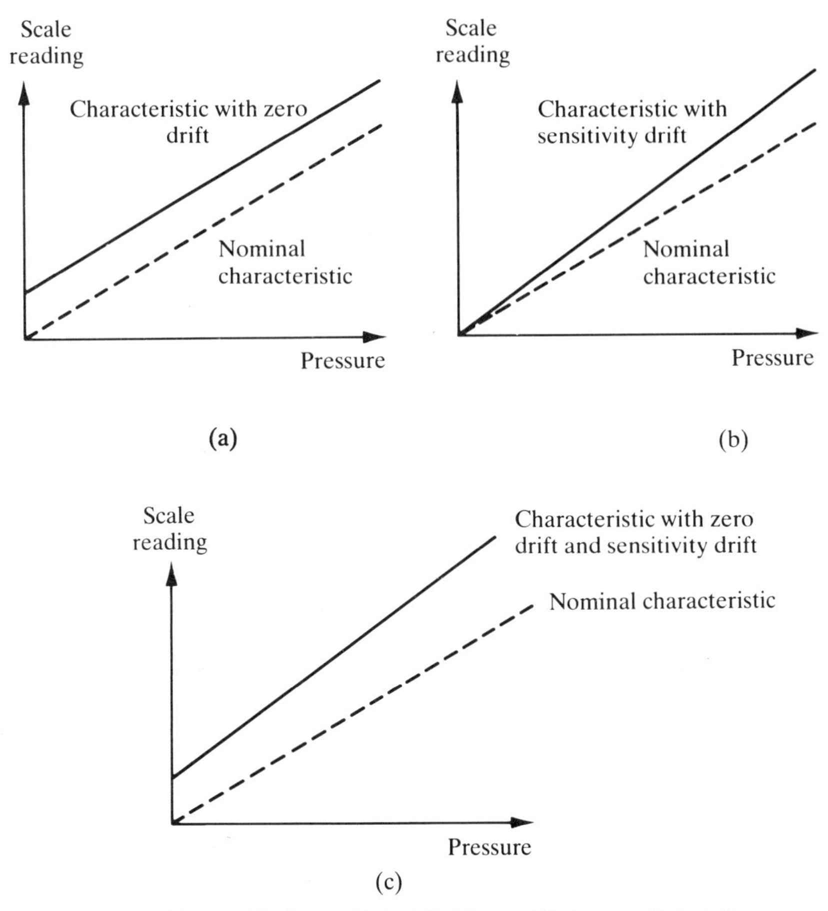

Figure 2.3 *(a) zero drift; (b) sensitivity drift; (c) zero drift plus sensitivity drift*

EXAMPLE 2.2

A spring balance is calibrated in an environment at a temperature of 20 °C and has the following deflection/load characteristic:

Load (kg):	0	1	2	3
Deflection (mm):	0	20	40	60

It is then used in an environment at a temperature of 30 °C and the following deflection/load characteristic is measured:

Load (kg):	0	1	2	3
Deflection (mm):	5	27	49	71

Determine the zero drift and sensitivity drift per °C change in ambiant temperature.

SOLUTION

At 20 °C, deflection/load characteristic is a straight line. Sensitivity = 20 mm/kg.

At 30 °C, deflection/load characteristic is still a straight line. Sensitivity = 22 mm/kg.

Bias (zero drift) = 5 mm (the no-load deflection)
Sensitivity drift = 2 mm/kg
Zero drift/°C = 5/10 = 0.5 mm/°C
Sensitivity drift/°C = 2/10 = 0.2 (mm per kg)/°C

2.2.9 Hysteresis

Figure 2.4 illustrates the output characteristic of an instrument which exhibits hysteresis. If the input measured quantity to the instrument is steadily increased from a negative value, the output reading varies in the manner shown in curve (A). If the input variable is then steadily decreased, the output varies in the manner shown in curve (B). The non-coincidence between these loading and unloading curves is known as hysteresis.

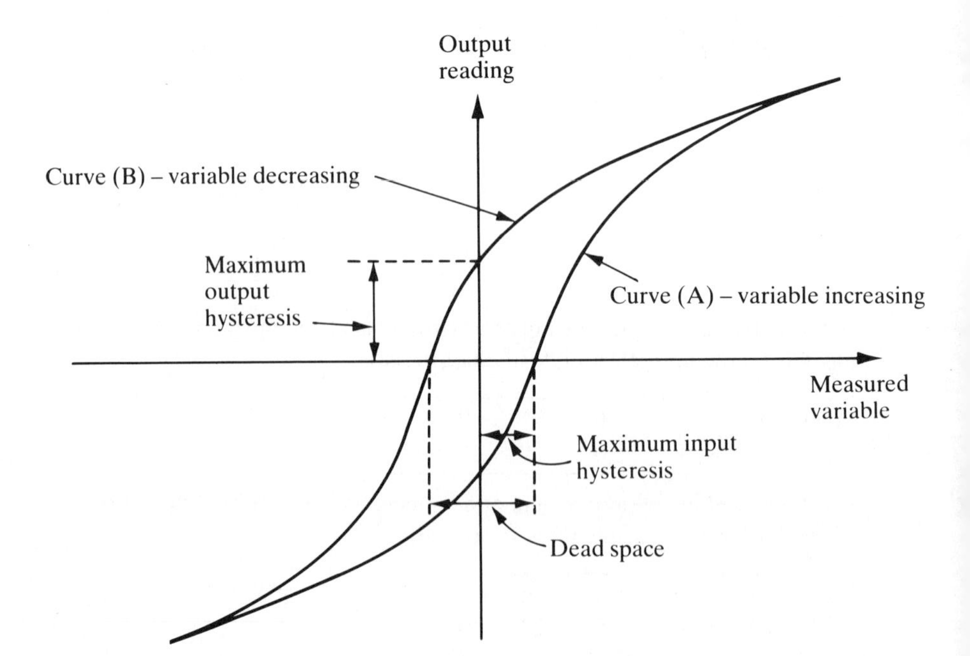

Figure 2.4 *Instrument characteristic with hysteresis*

Two quantities, maximum input hysteresis and maximum output hysteresis, are defined as shown in Figure 2.4 . These are normally expressed as a percentage of the full-scale input or output reading respectively.

2.2.10 **Dead space**

Dead space is defined as the range of different input values over which there is no change in output value.

Any instrument which exhibits hysteresis also displays dead space, as marked on Figure 2.4. Some instruments which do not suffer from any significant hysteresis can still exhibit a dead space in their output characteristics, however. Backlash in gears is a typical cause of dead space, and results in the sort of instrument output characteristic shown in Figure 2.5.

2.2.11 **Threshold**

If the input to an instrument is gradually increased from zero, the input will have to reach a certain minimum level before the change in the instrument output reading is of a large enough magnitude to be detectable. This minimum level of input is known as the threshold of the instrument. Manufacturers vary in the way that they specify threshold for instruments. Some quote absolute values, whereas others quote threshold as a percentage of full-scale readings.

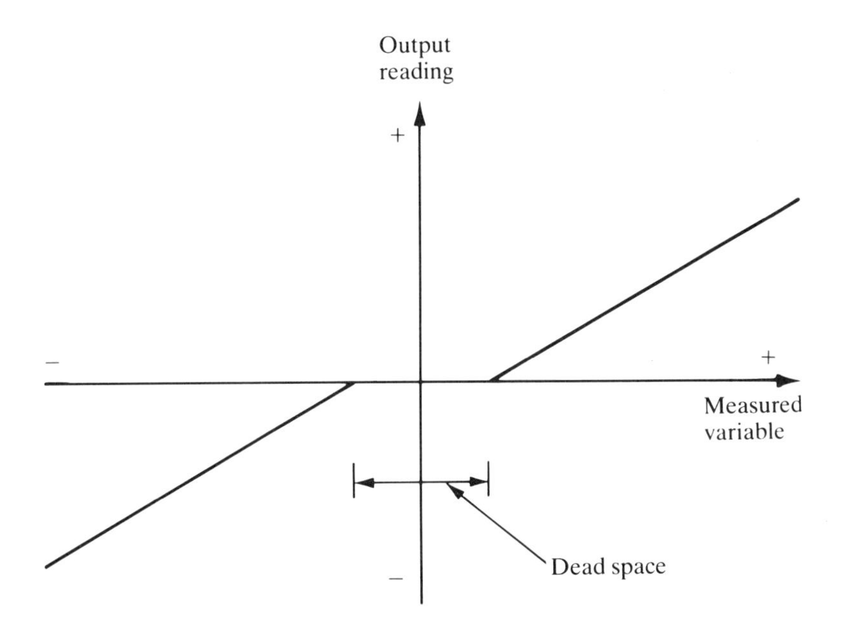

Figure 2.5 *Instrument characteristic with dead space*

As an illustration, a car speedometer typically has a threshold of about 15 km/h. This means that, if the vehicle starts from rest and accelerates, no output reading is observed on the speedometer until the speed reaches 15 km/h.

2.2.12 **Resolution**

When an instrument is showing a particular output reading, there is a lower limit on the magnitude of the change in the input measured quantity which produces an observable change in the instrument output. Like threshold, resolution is sometimes specified as an absolute value and sometimes as a percentage of f.s. deflection.

One of the major factors influencing the resolution of an instrument is how finely its output scale is divided into subdivisions. Using a car speedometer as an example again, this has subdivisions of, typically, 20 km/h. This means that when the needle is between the scale markings, we cannot estimate speed more accurately than to the nearest 5 km/h. This figure of 5 km/h thus represents the resolution of the instrument.

2.3 **Dynamic characteristics**

The static characteristics of measuring instruments are concerned only with the steady-state reading that the instrument settles down to, such as the accuracy of the reading.

The dynamic characteristics of a measuring instrument describe its behaviour between the time a measured quantity changes value and the time when the instrument output attains a steady value in response. As with static characteristics, any values for dynamic characteristics quoted in instrument data sheets only apply when the instrument is used under specified environmental conditions. Outside these calibration conditions, some variation in the dynamic parameters can be expected.

In any linear, time-invariant measuring system, the following general relation can be written between input and output for time $(t) > 0$:

$$a_n \frac{\mathrm{d}_n q_o}{\mathrm{d}t_n} + a_{n-1} \frac{\mathrm{d}_{n-1} q_o}{\mathrm{d}t_{n-1}} + \ldots + a_1 \frac{\mathrm{d}q_o}{\mathrm{d}t} + a_0 q_o$$

$$= b_m \frac{\mathrm{d}_m q_i}{\mathrm{d}t_m} + b_{m-1} \frac{\mathrm{d}_{m-1} q_i}{\mathrm{d}t_{m-1}} + \ldots + b_1 \frac{\mathrm{d}q_i}{\mathrm{d}t} + b_0 q_i \qquad (2.1)$$

where q_i is the measured quantity, q_o is the output reading and $a_0 \ldots a_n, b_0 \ldots b_m$ are constants.

The reader whose mathematical background is such that the above equation appears daunting should not worry unduly, as only certain special, simplified cases of it are applicable in normal measurement situations. The major point of importance is to have a practical appreciation of the manner in which various different types of instrument respond when the measurand applied to them varies.

If we limit consideration to that of step changes in the measured quantity only, then equation (2.1) reduces to:

$$a_n \frac{\mathrm{d}_n q_o}{\mathrm{d}t_n} + a_{n-1} \frac{\mathrm{d}_{n-1} q_o}{\mathrm{d}t_{n-1}} + \ldots + a_1 \frac{\mathrm{d}q_o}{\mathrm{d}t} + a_0 q_o = b_0 q_i \tag{2.2}$$

Further simplification can be made by taking certain special cases of equation (2.2) which collectively apply to nearly all measurement systems.

2.3.1 *Zero order instrument*

If all the coefficients $a_1 \ldots a_n$ other than a_0 in equation (2.2) are assumed zero, then:

$$a_0 q_o = b_0 q_i \quad \text{or} \quad q_0 = \frac{b_0 q_i}{a_0} = K q_i \tag{2.3}$$

where K is a constant known as the instrument sensitivity as defined earlier.

Any instrument which behaves according to equation (2.3) is said to be of zero order type. A potentiometer, which measures motion, is a good example of such an instrument, where the output voltage changes instantaneously as the slider is displaced along the potentiometer track.

2.3.2 *First order instrument*

If all the coefficients $a_2 \ldots a_n$ except for a_0 and a_1 are assumed zero in equation (2.2) then:

$$a_1 \frac{\mathrm{d}q_o}{\mathrm{d}t} + a_0 q_o = b_0 q_i \tag{2.4}$$

Any instrument which behaves according to equation (2.4) is known as a first order instrument.

Replacing d/dt by the D operator ($D \equiv \mathrm{d}/\mathrm{d}t$) in equation (2.4), we get:

$$a_1 D q_o + a_0 q_o = b_0 q_i$$

and rearranging:

$$q_o = \frac{(b_0/a_0) q_i}{[1 + (a_1/a_0)D]} \tag{2.5}$$

Defining $K = b_0/a_0$ as the static sensitivity and $\tau = a_1/a_0$ as the time constant of the system, equation (2.5) becomes:

$$q_o = \frac{K q_i}{1 + \tau D} \tag{2.6}$$

If equation (2.6) is solved analytically, the output quantity q_o in response to a step change in q_i varies with time in the manner shown in Figure 2.6. The time constant τ of

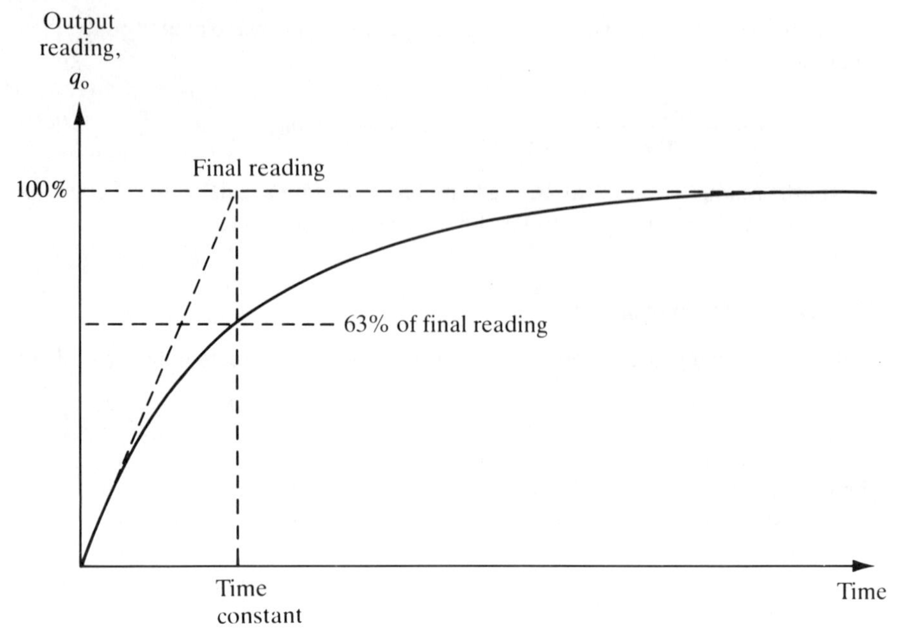

Figure 2.6 *Step response characteristic of first order instrument*

the step response is the time taken for the output quantity q_o to reach 63% of its final value.

The thermometer (see Chapter 9) is a good example of a first order instrument. It is well known that, if a thermometer at room temperature is plunged into boiling water, the output electromotive force (e.m.f.) does not rise instantaneously to a level indicating 100 °C, but instead approaches a reading indicating 100 °C in a manner similar to that shown in Figure 2.6.

A large number of other instruments also belong to this first order class: this is of particular importance in control systems where it is necessary to take account of the time lag that occurs between a measured quantity changing in value and the measuring instrument indicating the change. Fortunately, the time constant of many first order instruments is small relative to the dynamics of the process being measured, and so no serious problems are created.

EXAMPLE 2.3

A balloon is equipped with temperature and altitude measuring instruments and has radio equipment which can transmit the output readings of these instruments back to ground. The balloon is initially anchored to the ground with the instrument output readings in steady state. The altitude measuring instrument is approximately zero order and the temperature transducer first order with a time constant of 15 seconds. The temperature on the ground, T_0, is 10 °C and the temperature T_x at an altitude of x metres is given by the relation:

$$T_x = T_0 - 0.01x.$$

(a) If the balloon is released at time zero, and thereafter rises upwards at a velocity of 5 metres/second, draw a table showing the temperature and altitude measurements reported at intervals of 10 seconds over the first 50 seconds of travel. Show also in the table the error in each temperature reading.
(b) What temperature does the balloon report at an altitude of 5000 metres?

SOLUTION

This question requires the analytic solution of a first order differential equation and it is assumed that the reader has covered the principles of this previously in a mathematics course. If the reader is not familiar with this type of mathematical working, he or she may need assistance to work through this solution. The solution uses the *D*-operator and proceeds as follows. (NB a solution can also be obtained by using the Laplace transform and standard Laplace transform tables which some readers may prefer.)

Let the temperature reported by the balloon at some general time t be T_r. Then T_x is related to T_r by the relation:

$$T_r = \frac{T_x}{1+\tau D} = \frac{T_0 - 0.01x}{1+\tau D} = \frac{10 - 0.01x}{1+15D}$$

It is given that $x = 5t$. Thus:

$$T_r = \frac{10 - 0.05t}{1+15D}$$

The transient or complementary function part of the solution ($T_x = 0$) is given by:

$$T_{r_{cf}} = Ce^{-t/15}$$

where C is a constant derived from the initial conditions. The particular integral part of the solution is given by:

$$T_{r_{pi}} = 10 - 0.05(t - 15)$$

Thus the whole equation is given by:

$$T_r = T_{r_{cf}} + T_{r_{pi}} = Ce^{-t/15} + 10 - 0.05(t - 15)$$

Applying initial conditions: At $t = 0$, $T_r = 10$, i.e.

$$10 = Ce^{-0} + 10 - 0.05(-15)$$

Thus $C = -0.75$ and the solution can be written as:

$$T_r = 10 - 0.75e^{-t/15} - 0.05(t - 15)$$

Table 2.1

Times (s)	Altitude (m)	Temperature reading (°C)	Temperature error (°C)
0	0	10	0
10	50	9.86	0.36
20	100	9.55	0.55
30	150	9.15	0.65
40	200	8.70	0.70
50	250	8.22	0.72

(a) Using the above expression to calculate T_r for various values of t, Table 2.1 can be constructed.
(b) At 5000 m, $t = 1000$ seconds. Calculating T_r from the above expression:

$$T_r = 10 - 0.75e^{-1000/15} - 0.05(1000 - 15)$$

The exponential term approximates to zero and so T_r can be written as:

$$T_r \approx 10 - 0.05(985) = -39.25°C$$

This result might have been inferred from the table above where it can be seen that the error is converging towards a value of 0.75. For large values of t, the transducer reading lags the true temperature value by a period of time equal to the time constant of 15 seconds. In this time, the balloon travels a distance of 75 metres and the temperature falls by 0.75 °C. Thus for large values of t, the output reading is always 0.75 °C less than it should be.

2.3.3 *Second order instrument*

If all coefficients $a_3 ... a_n$ other than a_0, a_1 and a_2 in equation (2.2) are assumed zero, then we get:

$$a_2 \frac{d^2 q_o}{dt^2} + a_1 \frac{dq_o}{dt} + a_0 q_o = b_0 q_i \tag{2.7}$$

Applying the D operator again:

$$a_2 D^2 q_o + a_1 D q_o + a_0 q_o = b_0 q_i$$

and rearranging:

$$q_o = \frac{b_0 q_i}{a_0 + a_1 D + a_2 D^2} \tag{2.8}$$

It is convenient to re-express the variables a_0, a_1, a_2 and b_0 in equation (2.8) in terms of three parameters K (static sensitivity), ω (undamped natural frequency) and ζ (damping ratio), where

$$K = b_0/a_0, \quad \omega = \sqrt{a_0/a_2} \quad \text{and} \quad \zeta = \frac{a_1}{2\sqrt{a_0 a_2}}$$

Re-expressing equation (2.8) in terms of K, ω and ζ we get:

$$\frac{q_o}{q_i} = \frac{K}{D^2/\omega^2 + 2\zeta D/\omega + 1} \tag{2.9}$$

This is the standard equation for a second order system and any instrument whose response can be described by it is known as a second order instrument.

If equation (2.9) is solved analytically for a step input signal, the shape of the response obtained depends on the value of the damping ratio parameter ζ. The output step responses of a second order instrument for various values of ζ are shown in Figure 2.7 . For case (a) where $\zeta = 0$, there is no damping and the instrument output exhibits constant amplitude oscillations when disturbed by any change in the physical quantity measured. For light damping of $\zeta = 0.2$, represented by case (b), the response to a step change in input is still oscillatory but the oscillations gradually die down. Further increase in the value of ζ reduces oscillations and overshoot still more, as shown by curves (c) and (d), and finally the response becomes very overdamped as shown by curve (e) where the output reading creeps up slowly towards the correct reading. Clearly, the extreme response curves (a) and (e) are grossly unsuitable for any measuring instrument.

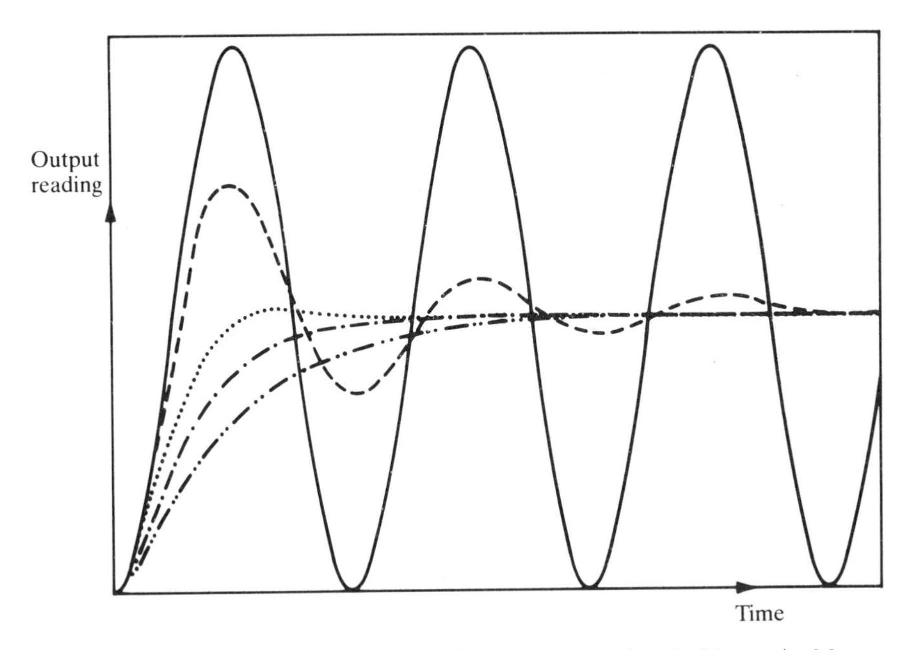

Figure 2.7 *Step responses of second order instruments: (a) ___ $\zeta = 0.0$; (b) - - - $\zeta = 0.2$; (c) ... $\zeta = 0.707$; (d) _ . . _ $\zeta = 1.0$; (e) _ . . . _ $\zeta = 1.5$*

If an instrument were to be only ever subjected to step inputs, then the design strategy would be to aim towards a damping ratio of 0.707, which gives the critically damped response (c). Unfortunately, most of the physical quantities which instruments are required to measure do not change in the mathematically convenient form of steps, but rather in the form of ramps of varying slopes. As the form of the input variable changes, so the best value for ζ varies, and choice of ζ becomes one of compromise between those values that are best for each type of input variable behaviour anticipated. Commercial second order instruments, of which the accelerometer is a common example, are generally designed to have a damping ratio (ζ) somewhere in the range of 0.6–0.8.

2.4 Calibration

The foregoing discussion has described the static and dynamic characteristics of measuring instruments in some detail. An important qualification which has been omitted from this discussion, however, is the statement that an instrument only conforms to stated static and dynamic patterns of behaviour after it has been calibrated. It can normally be assumed that a new instrument will have been calibrated when it is obtained from an instrument manufacturer, and will therefore initially behave according to the characteristics stated in the specifications. During use, however, its behaviour will gradually diverge from the stated specification for a variety of reasons. Such reasons include mechanical wear, and the effects of dirt, dust, fumes and chemicals in the operating environment. The rate of divergence from standard specifications varies according to the type of instrument, the frequency of usage and the severity of the operating conditions. However, there will come a time, determined by practical knowledge, when the characteristics of the instrument will have drifted from the standard specification by an unacceptable amount.

When this situation is reached, it is necessary to recalibrate the instrument back to the standard specifications. Such recalibration is performed by adjusting the instrument at each point in its output range until its output readings are the same as those of a second standard instrument to which the same inputs are applied. This second instrument is one kept solely for calibration purposes whose specifications are accurately known. Calibration procedures are discussed more fully in Chapter 4.

2.5 Choice of instruments

The first rule about choosing instruments for a particular measurement application is to match the instrument to the requirements. In particular, it is pointless to choose an instrument whose performance is much better (in terms of accuracy, etc.) than required in a particular measurement situation, as this will incur an unnecessary cost penalty with no tangible gain. For instance, if we have a thermometer in a room and its reading shows a temperature of 20 °C, then it does not really matter whether the true temperature of the room is 19.5 or 20.5 °C. Such small variations around 20 °C are too small to affect whether we feel warm enough or not. Our bodies cannot discriminate between such close levels of temperature and therefore a thermometer with an inaccuracy of ±0.5 °C is perfectly adequate. However, if we had to measure the temperature of certain chemical processes, a variation of 0.5 °C might have a significant effect on the rate of reaction or even the products of a process. A measurement inaccuracy that is much less than ±0.5 °C is therefore clearly required in that case.

Therefore, the starting point in choosing the most suitable instrument to use for measurement of a particular quantity in a manufacturing plant or other system is to study the specification of the instrument characteristics required carefully, especially parameters such as the desired measurement accuracy, resolution and sensitivity. It is also essential to know the environmental conditions that the instrument will be subjected to, as certain conditions will immediately eliminate the possibility of using some types of instrument. Provision of this type of information usually requires the expert knowledge of personnel who are intimately acquainted with the operation of the manufacturing plant or system in question. Then, a skilled instrument engineer, who has knowledge of all the instruments that are available for measuring the quantity in question, will be able to evaluate the possible list of instruments in terms of their accuracy, cost and suitability for the enviromental conditions, and thus choose the most appropriate instrument. As far as possible, measurement systems and instruments should be chosen which are as insensitive as possible to the operating environment, although this requirement is often difficult to meet because of cost and other performance considerations.

Published literature is of considerable help in the choice of a suitable instrument for a particular measurement situation. Many books are available which give valuable assistance in the necessary evaluation by providing lists and data about all the instruments available for measuring a range of physical quantities (e.g. Morris, 1993). However, new techniques and instruments are being developed all the time, and therefore a good instrumentation engineer must keep abreast of the latest developments by reading the appropriate technical journals regularly.

The instrument characteristics discussed earlier in this chapter are those features that form the technical basis for a comparison between the relative merits of different instruments. Generally, the better the characteristics, the higher the cost. However, in comparing the cost and relative suitability of different instruments for a particular measurement situation, considerations of durability, maintainability and constancy of performance are also very important because the instrument chosen will often have to be capable of operating for long periods without performance degradation and a requirement for costly maintenance. In consequence of this, the initial cost of an instrument often has a low weighting in the evaluation exercise.

Cost is very strongly correlated with the performance of an instrument, as measured by its static characteristics. Increasing the accuracy or resolution of an instrument, for example, can only be done at a penalty of increasing its manufacturing cost. Instrument choice therefore proceeds by specifying the minimum characteristics required by a measurement situation and then searching manufacturers' catalogs to find an instrument whose characteristics match those required. To select an instrument with characteristics superior to those required would only mean paying more than necessary for a level of performance greater than that needed.

As well as purchase cost, other important factors in the assessment exercise are instrument durability and the maintenance requirements. Assuming that one had £10 000 to spend, one would not spend £8000 on a new motor car whose projected life was five years if a car of equivalent specification with a projected life of ten years was available for £10 000. Likewise, durability is an important consideration in the choice of instruments. The projected life of instruments often depends on the conditions in which the instrument will have to operate. Maintenance requirements must also be taken into account, as they also have cost implications.

As a general rule, a good assessment criterion is obtained if the total purchase cost and estimated maintenance costs of an instrument over its life are divided by the period of its expected life. The figure obtained is thus a cost per year. However, this rule becomes modified where instruments are being installed on a process whose life is expected to be limited, perhaps in the manufacture of a particular model of car. Then, the total costs can only be divided by the period of time an instrument is expected to be used, unless an alternative use for the instrument is envisaged at the end of this period.

To summarize therefore, instrument choice is a compromise between performance characteristics, ruggedness and durability, maintenance requirements and purchase cost. To carry out such an evaluation properly, the instrument engineer must have a wide knowledge of the range of instruments available for measuring particular physical quantities, and he or she must also have a deep understanding of how instrument characteristics are affected by particular measurement situations and operating conditions.

2.6 Self-assessment questions

2.1 Briefly define and explain all the static characteristics of measuring instruments.

2.2 Explain the difference between accuracy and precision in an instrument.

2.3 A tungsten/5% rhenium–tungsten/26% rhenium thermocouple has an output e.m.f. as shown in the following table when its hot (measuring) junction is at the temperatures shown. Determine the sensitivity of measurement for the thermocouple in mV/°C.

mV:	4.37	8.74	13.11	17.48
°C:	250	500	750	1000

2.4 Define sensitivity drift and zero drift. What factors can cause sensitivity drift and zero drift in instrument characteristics?

2.5 (a) An instrument is calibrated in an environment at a temperature of 20 °C and the following output readings y are obtained for various input values x:

y:	13.1	26.2	39.3	52.4	65.5	78.6
x:	5	10	15	20	25	30

Determine the measurement sensitivity, expressed as the ratio y/x.
(b) When the instrument is subsequently used in an environment at a temperature of 50 °C, the input/output characteristic changes to the following:

y:	14.7	29.4	44.1	58.8	73.5	88.2
x:	5	10	15	20	25	30

Determine the new measurement sensitivity. Hence determine the sensitivity drift due to the change in ambiant temperature of 30 °C.

2.6 A load cell is calibrated in an environment at a temperature of 21 °C and has the following deflection/load characteristic:

Load (kg):	0	50	100	150	200
Deflection (mm):	0.0	1.0	2.0	3.0	4.0

When used in an environment at 35 °C, its characteristic changes to the following:

Load (kg):	0	50	100	150	200
Deflection (mm):	0.2	1.3	2.4	3.5	4.6

(a) Determine the sensitivity at 21 °C and 35 °C.
(b) Calculate the total zero drift and sensitivity drift at 35 °C.
(c) Hence determine the zero drift and sensitivity drift coefficients (in units of μm/°C and (μm per kg)/(°C).

2.7 An unmanned submarine is equipped with temperature and depth measuring instruments and has radio equipment which can transmit the output readings of these instruments back to the surface. The submarine is initially floating on the surface of the sea with the instrument output readings in steady state. The depth measuring instrument is approximately zero order and the temperature transducer first order with a time constant of 50 seconds. The water temperature on the sea surface, T_0, is 20 °C and the temperature T_x at a depth of x metres is given by the relation:

$$T_x = T_0 - 0.01x$$

(a) If the submarine starts diving at time zero, and thereafter goes down at a velocity of 0.5 metres/second, draw a table showing the temperature and depth measurements reported at intervals of 100 seconds over the first 500 seconds of travel. Show also in the table the error in each temperature reading.
(b) What temperature does the submarine report at a depth of 1000 metres?

2.8 Write down the general differential equation describing the dynamic response of a second order measuring instrument and state the expressions relating the static sensitivity, undamped natural frequency and damping ratio to the parameters in this differential equation. Sketch the instrument response for the cases of heavy damping, critical damping and light damping, and state which of these is the usual target when a second order instrument is being designed.

References and further reading

BS 5233, *Glossary of Terms used in Metrology* (incorporating BS 2643), 1986. British Standards Institution, London.

BS 5532, *Statistical Terminology*, 1978. British Standards Institution, London.

ISO 3534, *Statistics — Vocabulary and Symbols*, 1977. International Organization for Standards, Geneva.

Morris, A.S., *Principles of Measurement and Instrumentation*, 1993. Prentice Hall, Hemel Hempstead.

CHAPTER 3

Measurement errors

3.1 Introduction

This chapter is concerned with identifying the various errors which exist in a measurement system, and suggesting mechanisms for reducing their magnitude and effect. A discussion is also included about the way in which the separate error components are combined together in order to calculate the overall measurement system error level.

It is extremely important in any measurement system to reduce errors in instrument output readings to the minimum possible level and to quantify the maximum error which may exist in any output reading. A prerequisite in this is a detailed analysis of the sources of error that exist. Such errors in measurement data can be divided into two groups, known as systematic errors and random errors.

Systematic errors describe errors in the output readings of a measurement system which are consistently on one side of the correct reading, i.e. either all the errors are positive or they are all negative. Two major sources of systematic errors are system disturbance during measurement and the effect of modifying inputs, as discussed in sections 3.2.1 and 3.2.2. Other sources of systematic error include bent meter needles, the use of uncalibrated instruments, poor cabling practices and the thermal generation of electromotive forces (commonly known as thermal e.m.f.s). The latter two sources are considered in section 3.2.3. Even when systematic errors due to the above factors have been reduced or eliminated, some errors remain which are inherent in the manufacture of an instrument. These are quantified by the accuracy figure quoted in the published specifications contained in the instrument data sheet.

Random errors are perturbations of the measurement either side of the true value caused by random and unpredictable effects, such that positive errors and negative errors occur in approximately equal numbers for a series of measurements made of the same quantity. Such perturbations are mainly small, but large perturbations occur from time to time, again unpredictably. Random errors often arise when measurements are taken by human observation of an analog meter, especially where this involves interpolation between scale points. Electrical noise can also be a source of random errors. To a large extent, random errors can be overcome by taking the same measurement a number of times and extracting a value by averaging or other statistical techniques, as discussed in section 3.3.1. However, any quantification of the measurement value and statement of error bounds remains a statistical quantity.

Because of the nature of random errors and the fact that large perturbations in the measured quantity occur from time to time, the best that we can do is to express measurements in probabilistic terms: we may be able to assign a 95% or even 99% confidence level that the measurement is a certain value within error bounds of, say, ±1%, but we can never attach a 100% probability to measurement values which are subject to random errors.

Finally, a word must be said about the distinction between systematic and random errors. Error sources in the measurement system must be examined carefully to determine what type of error is present, systematic or random, and to apply the appropriate treatment. In the case of manual data measurements, a human observer may make a different observation at each attempt, but it is often reasonable to assume that the errors are random and that the mean of these readings is likely to be close to the correct value. However, this is only true as long as the human observer is not introducing a systematic parallax-induced error as well by persistently reading the position of a needle against the scale of an analog meter from the same side rather than from directly above. In that case, correction would have to be made for this systematic error, or bias, in the measurements before statistical techniques were applied to reduce the effect of random errors.

3.2 Systematic errors

Systematic errors in the output of many instruments are due to factors inherent in the manufacture of the instrument arising out of tolerances in the components of the instrument. They can also arise due to wear in instrument components over a period of time. In other cases, systematic errors are introduced by either the effect of environmental disturbances or through the disturbance of the measured system by the act of measurement. These various sources of systematic error, and ways in which the magnitude of the errors can be reduced, are discussed below.

3.2.1 System disturbance due to measurement

Disturbance of the measured system by the act of measurement is one source of systematic error. If we were to start with a beaker of hot water and wished to measure its temperature with a mercury-in-glass thermometer, then we should take the thermometer, which would be initially at room temperature, and plunge it into the water. In so doing, we would be introducing the relatively cold mass of the thermometer into the hot water and a heat transfer would take place between the water and the thermometer. This heat transfer would lower the temperature of the water. While in this case the reduction in temperature would be so small as to be undetectable by the limited measurement resolution of such a thermometer, the effect is finite and clearly establishes the principle that, in nearly all measurement situations, the process of measurement disturbs the system and alters the values of the physical quantities being measured.

Another example is that of measuring car tyre pressures with the type of pressure gauge commonly obtainable from car accessory shops. Measurement is made by pushing one end of the pressure gauge onto the valve of the tyre and reading the displacement of the other end of the gauge against a scale. As the gauge is used, a quantity of air flows from the tyre into the gauge. This air does not subsequently flow back into the tyre after measurement, and so the tyre has been disturbed and the air pressure inside it has been permanently reduced.

Thus, as a general rule, the process of measurement always disturbs the system being measured. The magnitude of the disturbance varies from one measurement system to the next and is affected particularly by the type of instrument used for measurement. Ways of minimizing disturbance of measured systems is an important consideration in instrument design. A prerequisite for this, however, is a full understanding of the mechanisms of system disturbance.

Measurements in electric circuits are particularly prone to errors induced through the loading effect on the circuit when instruments are applied to make voltage and current measurements. For most electrical networks, circuit analysis methods such as Thevenin's theorem are needed to analyze such loading effects, and these are covered in more advanced texts (e.g. Morris, 1993). However, for the simple circuit shown in Figure 3.1, the analysis is fairly easy.

In this circuit, the voltage across the resistor with resistance R_2 is to be measured by a voltmeter with resistance R_m. Here, R_m acts as a shunt resistance across R_2, decreasing the resistance between points A and B and so disturbing the circuit. The voltage E_m measured by the meter is therefore not the value of the voltage E_0 that existed prior to measurement. The extent of the disturbance can be asssessed by calculating the open-circuit voltage E_0 and comparing it with E_m.

Starting with the unloaded circuit in Figure 3.1, the current I is given by Ohm's law as:

$$I = \frac{V}{R_1 + R_2}$$

Again, using Ohm's law, voltage across AB is then given by:

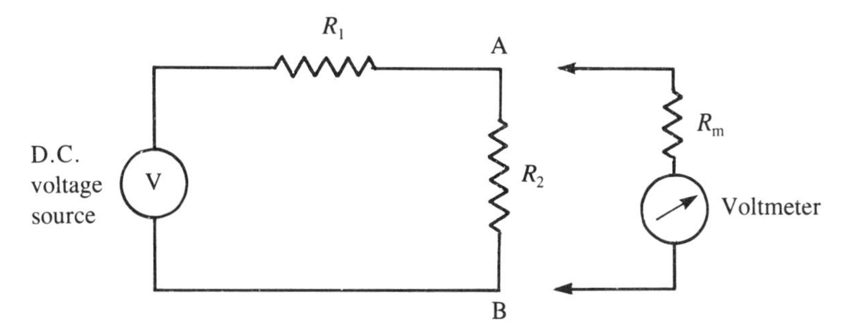

Figure 3.1 *Loading of circuit by adding voltmeter*

$$E_0 = IR_2 = \frac{VR_2}{R_1 + R_2} \tag{3.1}$$

With the voltmeter added to the circuit, there are now two resistances in parallel across AB, R_2 and R_m, and the expression for the resistance across AB can be written as:

$$R_{AB} = \frac{R_2 R_m}{R_2 + R_m}$$

Then, replacing R_2 by R_{AB} in equation (3.1) above, the voltage E_m measured by the meter is given by:

$$\begin{aligned} E_m = \frac{VR_{AB}}{R_1 + R_{AB}} &= \frac{VR_2R_m}{(R_2 + R_m)} \cdot \frac{1}{R_1 + (R_2R_m/R_2 + R_m)} \\ &= \frac{VR_2R_m}{R_1R_2 + R_m(R_1 + R_2)} \end{aligned} \tag{3.2}$$

Thus, from equations (3.1) and (3.2):

$$\frac{E_m}{E_0} = \frac{VR_2R_m}{[R_1R_2 + R_m(R_1 + R_2)]} \cdot \frac{(R_1 + R_2)}{VR_2} = \frac{R_m(R_1 + R_2)}{R_1R_2 + R_m(R_1 + R_2)} \tag{3.3}$$

If R_m is very large compared with R_1 and R_2, then $R_m(R_1+R_2) >> R_1R_2$ and then the denominator of the equation (3.3) approaches $R_m(R_1+R_2)$ and E_m/E_0 approaches unity, and thus E_m approaches E_0.

It is thus obvious that as R_m gets larger, the ratio E_m/E_0 gets closer to unity, showing that the design strategy should be to make R_m as large as possible in order to minimize the disturbance of the measured system. (Note that we did not calculate the value of E_0, since this was not required in quantifying the effect of R_m.)

EXAMPLE 3.1

Suppose that the components of the circuit shown in Figure 3.1 have the following values: $R_1 = 500\,\Omega$; $R_2 = 500\,\Omega$. The voltage across AB is measured by a voltmeter whose internal resistance is 4750 Ω. What is the measurement error caused by the resistance of the measuring instrument?

SOLUTION

Proceeding by substituting the given component values into equation (3.3), we obtain:

$$\frac{E_m}{E_0} = \frac{4750 \times 1000}{[(25 \times 10^4) + (4750 \times 1000)]} = 0.95$$

Thus the error in the measured value is 5%.

At this point, it is interesting to note what constraints exist when practical attempts are made to achieve a high internal resistance in the design of a moving-coil voltmeter. Such an instrument consists of a coil carrying a pointer mounted in a fixed magnetic field. As current flows through the coil, the interaction between the field generated and the fixed field causes the pointer it carries to turn in proportion to the applied current (see Chapter 7, especially Figure 7.2, for further details).

The simplest way of increasing the input resistance of the meter, or as it is more commonly called, the input impedance, is either to increase the number of turns in the coil or to construct the same number of coil turns with a higher-resistance material. Either of these solutions decreases the current in the coil, however, giving less magnetic torque and thus decreasing the measurement sensitivity of the instrument (i.e. for a given applied voltage, we get less deflection of the pointer). This problem can be overcome by changing the spring constant of the restraining springs of the instrument, such that less torque is required to turn the pointer by a given amount. This, however, reduces the ruggedness of the instrument and also demands better pivot design to reduce friction. This highlights a very important but tiresome principle in instrument design: any attempt to improve the performance of an instrument in one aspect generally decreases the performance in some other aspect. This is an inescapable fact of life with passive instruments such as the type of voltmeter mentioned, and is often the reason for the use of alternative active instruments such as digital voltmeters, where the inclusion of auxiliary power greatly improves performance.

3.2.2 Modifying inputs in measurement systems

The fact that the static and dynamic characteristics of measuring instruments are specified for particular environmental conditions, for example, temperature and pressure, has already been discussed at considerable length in Chapter 2. These specified conditions must be reproduced as closely as possible during calibration exercises because, away from the specified calibration conditions, the characteristics of measuring instruments vary to some extent. The magnitude of this variation is quantified by the two constants known as the sensitivity drift and the zero drift, defined previously in Chapter 2, both of which are generally included in the published specifications for an instrument. Such variations of environmental conditions away from the calibration conditions are described as modifying inputs to the system and are a further source of systematic error. The environmental variation is described as a measurement system input because the effect on the system output is the same as if the value of the measured quantity, which is the real input, had changed by a certain amount.

Without proper analysis, it is impossible to establish how much of an instrument's output is due to the real input and how much is due to one or more modifying inputs. This is illustrated by the following example.

Suppose that we have a small closed box weighing 0.1 kg when empty, which we think contains either a rat or a mouse. If we put the box on the bathroom scales and observe a reading of 1.0 kg, this does not immediately tell us what is in the box because the reading may be due to one of three things:

1 A 0.9 kg rat in the box (real input).
2 An empty box with a 0.9 kg bias on the scales due to a temperature change (modifying input).
3 A 0.4 kg mouse in the box together with a 0.5 kg bias (real plus modifying inputs).

Thus, the magnitude of any modifying input must be measured before the value of the measured quantity, which is the real input, can be determined from the output reading of an instrument.

In any general measurement situation, it is very difficult to avoid modifying inputs, because it is either impractical or impossible to control the environmental conditions surrounding the measurement system. System designers are therefore charged with the task of either reducing the susceptibility of measuring instruments to modifying inputs or alternatively quantifying the effect of modifying inputs and correcting for them in the instrument output reading.

The techniques used to deal with modifying inputs and to minimize their effect on the final output measurement follow a number of routes, as discussed below.

Careful instrument design

Careful instrument design is the most useful weapon in the battle against modifying inputs, by reducing the sensitivity of an instrument to modifying inputs to as low a level as possible. In the design of strain gauges for instance, the element should be constructed from a material whose resistance has a very low temperature coefficient (i.e. the variation of the resistance with temperature is very small). For many instruments, however, it is not possible to reduce their sensitivity to modifying inputs to a satisfactory level by simple design adjustments, and other measures have to be taken.

Method of opposing inputs

The method of opposing inputs compensates for the effect of a modifying input in a measurement system by introducing an equal and opposite modifying input which cancels it out. One example of how this technique is applied is in the type of millivoltmeter shown in Figure 3.2. This consists of a coil suspended in a fixed magnetic field produced by a permanent magnet. When an unknown voltage is applied to the coil, the magnetic field due to the current interacts with the fixed field and causes the coil (and a pointer attached to the coil) to turn. If the coil resistance is sensitive to temperature, then any modifying input to the system in the form of a temperature change will alter the value of the coil current for a given applied voltage and so alter the pointer output reading. Compensation for this is made by introducing a compensating resistance R_{comp} into the circuit, where R_{comp} has a temperature coefficient which is equal in magnitude but opposite in sign to that of the coil.

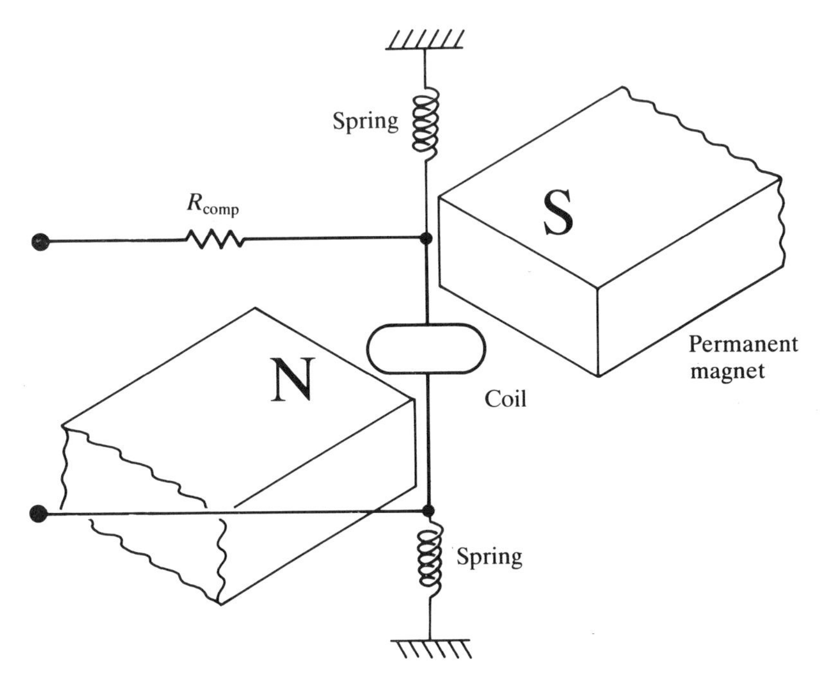

Figure 3.2 *Millivoltmeter*

High-gain feedback

The benefit of adding high-gain feedback to many measurement systems is illustrated by considering the case of the voltage measuring instrument whose block diagram is shown in Figure 3.3. In this system, the unknown voltage E_i is applied to a coil of torque constant K_c, and the torque induced turns a pointer against the restraining action of a spring with spring constant K_s. The effect of modifying inputs on the torque and spring constants are represented by variables D_c and D_s.

In the absence of modifying inputs, the displacement of the pointer X_o is given by:

$$X_o = K_c \cdot K_s \cdot E_i$$

However, in the presence of modifying inputs, both K_c and K_s change and the

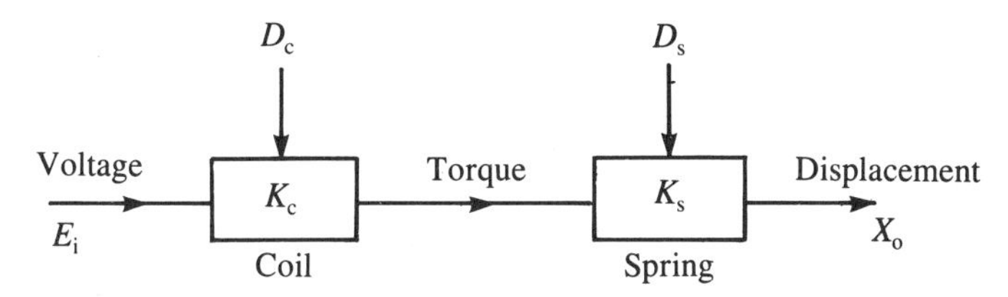

Figure 3.3 *Block diagram for voltage measuring instrument*

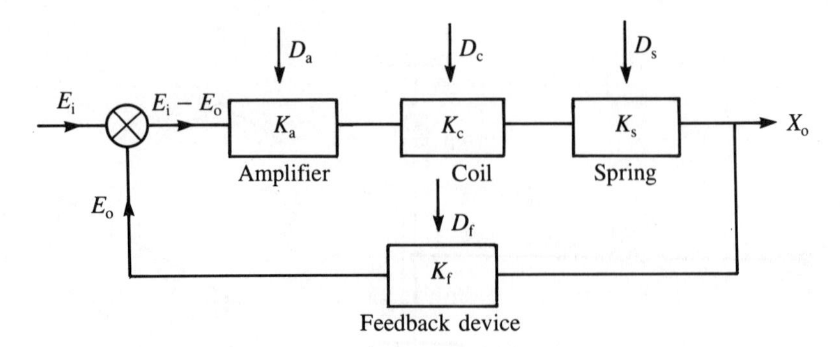

Figure 3.4 Block diagram of voltage measuring instrument with high-gain feedback

relationship between X_o and E_i can be affected greatly. It therefore becomes difficult or impossible to calculate E_i from the measured value of X_o.

Consider now what happens if the system is converted into a high-gain, closed-loop one, as shown in Figure 3.4, by adding an amplifier of gain constant K_a and a feedback device with gain constant K_f. Assume also that the effect of modifying inputs on the values of K_a and K_f are represented by D_a and D_f. The feedback device feeds back a voltage E_o proportional to the pointer displacement X_o. This is compared with the unknown voltage E_i by a comparator and the error is amplified.

Writing down the equations of the system, we have:

$$E_o = K_f \cdot X_o$$
$$X_o = (E_i - E_o) \cdot K_a \cdot K_c \cdot K_s = (E_i - K_f \cdot X_o) \cdot K_a \cdot K_c \cdot K_s$$

thus:

$$E_i \cdot K_a \cdot K_c \cdot K_s = (1 + K_f \cdot K_a \cdot K_c \cdot K_s) \cdot X_o$$

i.e.

$$X_o = \frac{K_a \cdot K_c \cdot K_s}{1 + K_f \cdot K_a \cdot K_c \cdot K_s} \cdot E_i \tag{3.4}$$

Because K_a is the gain constant of a high-gain amplifier, it is very large, and therefore $K_f \cdot K_a \cdot K_c \cdot K_s \gg 1$, and equation (3.4) can be reduced to:

$$X_o \approx \frac{E_i}{K_f}$$

This is a highly important result because we have reduced the relationship between X_o and E_i to one which involves only K_f. The sensitivity of the gain constants K_a, K_c and K_s to the modifying inputs D_a, D_c and D_s has thereby been rendered irrelevant and we only have to be concerned with one modifying input D_f. Conveniently, it is usually an easy matter to design a feedback device which is insensitive to modifying inputs: this is much easier than trying to make a coil or spring insensitive. Thus high-gain feedback techniques are often a very effective way of reducing a measurement system's sensitivity to modifying inputs. One potential problem which must be mentioned, however, is that there is a possibility that high-gain feedback will cause instability in the system. Any application of this method must therefore include careful stability analysis of the system.

Signal filtering

One frequent problem in measurement systems is corruption of the output reading by periodic noise, often at a frequency of 50 Hz caused by pick-up through the close proximity of the measurement system to apparatus or current-carrying cables operating on a mains supply. Periodic noise corruption at higher frequencies is also often introduced by mechanical oscillation or vibration within some component of a measurement system. The amplitude of all such noise components can be substantially attenuated by the inclusion of filtering of an appropriate form in the system, as discussed at greater length in Chapter 5. Band-stop filters can be especially useful where corruption is of one particular known frequency, or, more generally, low-pass filters are employed to attenuate all noise in the frequency range of 50 Hz and above.

Measurement systems with a low-level output, such as a bridge circuit measuring a strain-gauge resistance, are particularly prone to noise, and Figure 3.5(a) shows the typical corruption of a bridge output by 50 Hz pick-up. The beneficial effect of putting a simple passive RC low-pass filter across the output is shown in Figure 3.5(b).

3.2.3 *Other sources of systematic error*

Wear in instrument components

Systematic errors can frequently develop over a period of time because of wear in instrument components. Recalibration often provides a full solution to this problem.

Connecting leads

In connecting together the components of a measurement system, a common source of error is the failure to take proper account of the resistance of connecting leads, or pipes in the case of pneumatically or hydraulically actuated measurement systems. In typical applications of a resistance thermometer, for instance, it is common to find the thermometer separated from other parts of the measurement system by perhaps 30 metres. The resistance of such a length of 7/0.0076 copper wire is 2.5 Ω and there is a further complication that such wire has a temperature coefficient of 1 mΩ/°C.

Careful consideration therefore needs to be given to the choice of connecting leads. Not only should they be of adequate cross-section so that their resistance is minimized, but they should be adequately screened if they are thought likely to be subject to electrical or magnetic fields which could otherwise cause induced noise. Where screening is thought essential, then the routeing of cables also needs careful planning. In one application in the author's personal experience involving instrumentation of an electric-arc steelmaking furnace, screened signal-carrying cables between transducers on the arc furnace and a control room at the side of the furnace were initially

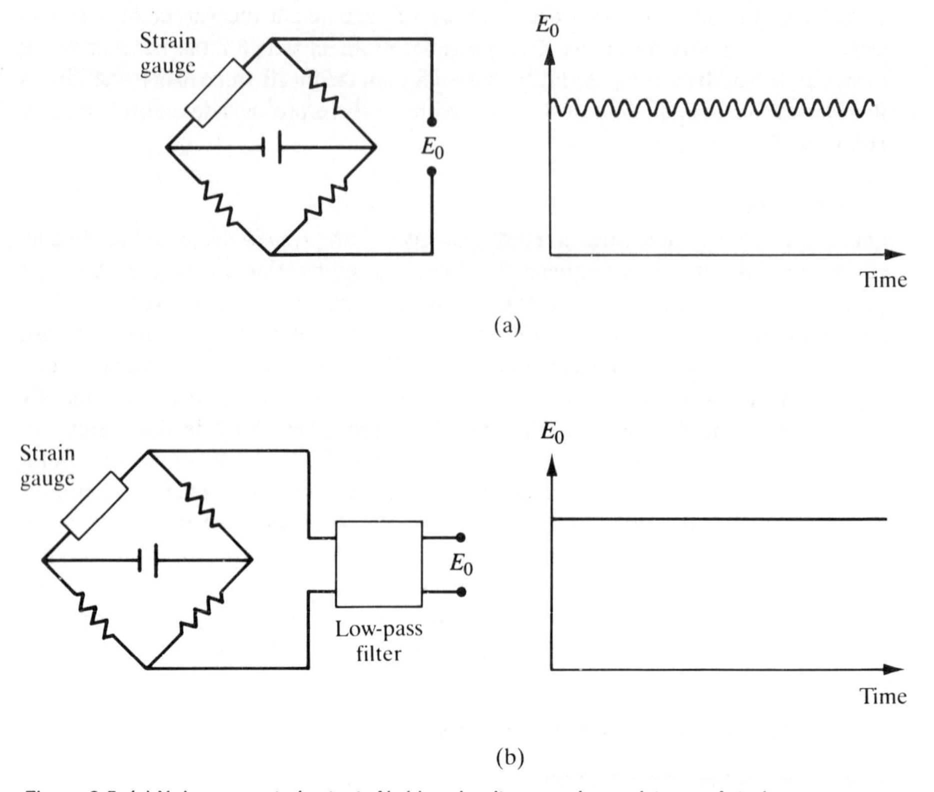

Figure 3.5 *(a) Noise-corrupted output of bridge circuit measuring resistance of strain gauge. (b) Effect of adding low-pass filter*

corrupted by high-amplitude 50 Hz noise. However, by changing the route of the cables between the transducers and the control room, the magnitude of this induced noise was reduced by a factor of about ten.

Thermal e.m.f.s

Whenever metals of two different types are connected together, a thermal e.m.f. is generated, which varies according to the temperature of the joint. This is known as the *thermoelectric effect* and is the physical principle on which temperature measuring thermocouples operate (see Chapter 9). Such thermal e.m.f.s are only a few millivolts in magnitude and so the effect is only significant when typical voltage output signals of a measurement system are of a similar low magnitude.

One such situation is where one e.m.f. measuring instrument is used to monitor the output of several thermocouples measuring the temperatures at different points in a process control system. This requires a means of automatically switching the output of each thermocouple to the measuring instrument in turn. Nickel–iron reed-relays with copper connecting leads are commonly used to provide this switching function. This introduces a thermocouple effect of magnitude 40 μV/°C between the reed-relay and the copper connecting leads. There is no problem if both ends of the reed relay are at the same temperature because then the thermal e.m.f.s will be equal and opposite and so cancel out. However, there are several recorded instances where, because of lack of awareness of the problem, poor design has resulted in the two ends of a reed-relay being at different temperatures and causing a net thermal e.m.f. The serious error that this introduces is clear. For a temperature difference between the two ends of only 2 °C, the thermal e.m.f. is 80 μV, which is very large compared with a typical thermocouple output level of 400 μV.

Another example of the difficulties that thermal e.m.f.s can create becomes apparent in considering the following problem which was reported in a current measuring system. This system had been designed such that the current in a particular part of a circuit was calculated by applying it to an accurately calibrated wire-wound resistance of value 100 Ω and measuring the voltage drop across the resistance. In calibration of the system, a known current of 20 μA was applied to the resistance and a voltage of 2.20 mV was measured by an accurate high-impedance instrument. Simple application of Ohm's law reveals that such a voltage reading indicates a current value of 22 μA. What then was the explanation for this discrepancy? The answer once again is a thermal e.m.f. Because the designer was not aware of thermal e.m.f.s, the circuit had been constructed such that one side of the standard resistance was close to a power transistor, creating a difference in temperature between the two ends of the resistor of 2 °C. The thermal e.m.f. associated with this was sufficient to account for the 10% measurement error found.

3.3 Random errors

Random errors in measurements are caused by random, unpredictable variations in the measurement system and they can largely be eliminated by calculating the mean or median of the measurements. The degree of confidence in the calculated mean/median values can be quantified by calculating the standard deviation or variance of the data, these being parameters which describe how the measurements are distributed about the mean/median value. All of these terms are explained more fully in Section 3.3.1.

Because of the unpredictability of random errors, any error bounds placed on measurements can only be quantified in probabilistic terms. Thus, if we say that the possible error in a measurement subject to random errors is ±2% of the measured value, we are only implying that this is probably true, i.e. there is, say, a 95% probability that the error level does not exceed ±2%.

The distribution of measurement data about the mean value can be displayed graphically by frequency distribution curves, as discussed in Section 3.3.2. Calculation of the area under the frequency distribution curve gives the probability that the error will lie between any two chosen error levels.

3.3.1 *Statistical analysis of data*

In the analysis of measurements subject to random errors, various parameters can be extracted. Formal definitions of these and their means of calculation are given in the following sections. It should be noted that 'hand calculation' of these parameters is rarely necessary nowadays, as many standard computer packages are available for their calculation, and this facility is also provided by many personal calculators and also intelligent instruments.

Mean and median values

In any measurement situation subject to random errors, the normal technique is to take the same reading a number of times, ideally using different observers, and extract the most likely value from the measurement data set. For a set of n measurements $x_1, x_2 \ldots x_n$, the most likely true value is the *mean* given by:

$$x_{\text{mean}} = \frac{x_1 + x_2 + \ldots + x_n}{n} \tag{3.5}$$

This is valid for all data sets where the measurement errors are distributed equally about the line of zero error, i.e. where the positive errors are balanced in quantity and magnitude by the negative errors.

When the number of values in the data set is large, however, calculation of the mean value is tedious, and it is more convenient to use the median value, this being a close approximation to the mean value. The *median* is given by the middle value when the measurements in the data set are written down in ascending order of magnitude.

For a set of n measurements $x_1, x_2, \ldots, x_n$ written down in ascending order of magnitude, the median value is given by:

$$x_{\text{median}} = x_{(n+1)/2}$$

Thus, for a set of nine measurements $x_1, x_2, \ldots, x_9$ the median value is x_5.

For an even number of data values, the median value is mid-way between the centre two values, i.e. for ten measurements $x_1 \ldots x_{10}$, the median value is given by: $(x_5 + x_6)/2$.

Suppose that, in a particular measurement situation, a mass is measured by a beam-balance, and the set of readings in grams shown in Table 3.1 is obtained at a particular time by different observers. The mean value of this set of data is 81.18 g, calculated according to equation (3.5). The median value is 81.1 g, which is the middle value if

Table 3.1 *Set of mass measurements subject to random errors*

81.6	81.1	81.2	81.0	81.3
81.1	80.5	81.8	81.3	81.6
81.4	81.3	81.1	81.1	81.1
80.9	80.8	81.5	80.8	

the data values are written down in ascending order, starting at 80.5 g and ending at 81.8 g.

Standard deviation and variance

The probability that the mean or median value of a data set represents the true measurement value depends on how widely scattered the data values are. If the values of mass measurements in Table 3.1 had ranged from 79 g up to 83 g, our confidence in the mean value would be much less. The spread of values about the mean is analyzed by first calculating the deviation of each value from the mean. For any general value x_i, the deviation d_i is given by:

$$d_i = x_i - x_{\text{mean}}$$

The extent to which n measurement values are spread about the mean can now be expressed by the standard deviation σ, where σ is given by:

$$\sigma = \sqrt{\frac{d_1{}^2 + d_2{}^2 + \ldots + d_n{}^2}{n-1}} \tag{3.6}$$

This spread can alternatively be expressed by the variance V, which is the square of the standard deviation, i.e. $V = \sigma^2$.

Mathematically minded readers may have observed that the expression for σ given above differs from the mathematical definition of the standard deviation, which has (n) instead of $(n - 1)$ in the denominator. This difference arises because the mathematical definition of σ is for an infinite data set, whereas in the case of measurements we are always concerned with finite data sets. For a finite set of measurements $(d_i)_{i=1,n}$, the mean x_m will differ from the true mean μ of the infinite data set that the finite set d_i is part of. If somehow we knew the true mean μ of a set of measurements, then the deviations d_i could be calculated as the deviation of each data value from the true mean, and it would be correct to calculate σ using (n) instead of $(n - 1)$ in the expression for σ in equation (3.6). However, in normal situations, using $(n - 1)$ in the denominator of equation (3.6) produces a value of the standard deviation which is statistically closer to the correct value.

EXAMPLE 3.2

The following measurements in mA were taken of the current in a circuit (the circuit was in steady state and therefore, although the measurements varied due to random errors, the current was actually constant):

21.5, 22.1, 21.3, 21.7, 22.0, 22.2, 21.8, 21.4, 21.9, 22.1

Calculate the mean value, the deviations from the mean and the standard deviation.

SOLUTION

Mean value = Σ(data values)/10 = 218/10 = 21.8 mA.

Now draw a table of measurements and deviations:

Measurement:	21.5	22.1	21.3	21.7	22.0	22.2	21.8	21.4	21.9	22.1
Deviation from mean:	−0.3	+0.3	−0.5	−0.1	+0.2	+0.4	0.0	−0.4	+0.1	+0.3
(Deviations)^2:	0.09	0.09	0.25	0.01	0.04	0.16	0.0	0.16	0.01	0.09

$\Sigma(\text{deviations})^2 = 0.90$
n = number of measurements = 10
$\Sigma(\text{deviations})^2/(n-1) = \Sigma(\text{deviations})^2/9 = 0.10$
$\sqrt{[\Sigma(\text{deviations})^2/9]} = 0.316$
Thus, standard deviation = 0.32 mA (the nature of the measurements do not justify expressing the standard deviation to any accuracy greater than two decimal places).

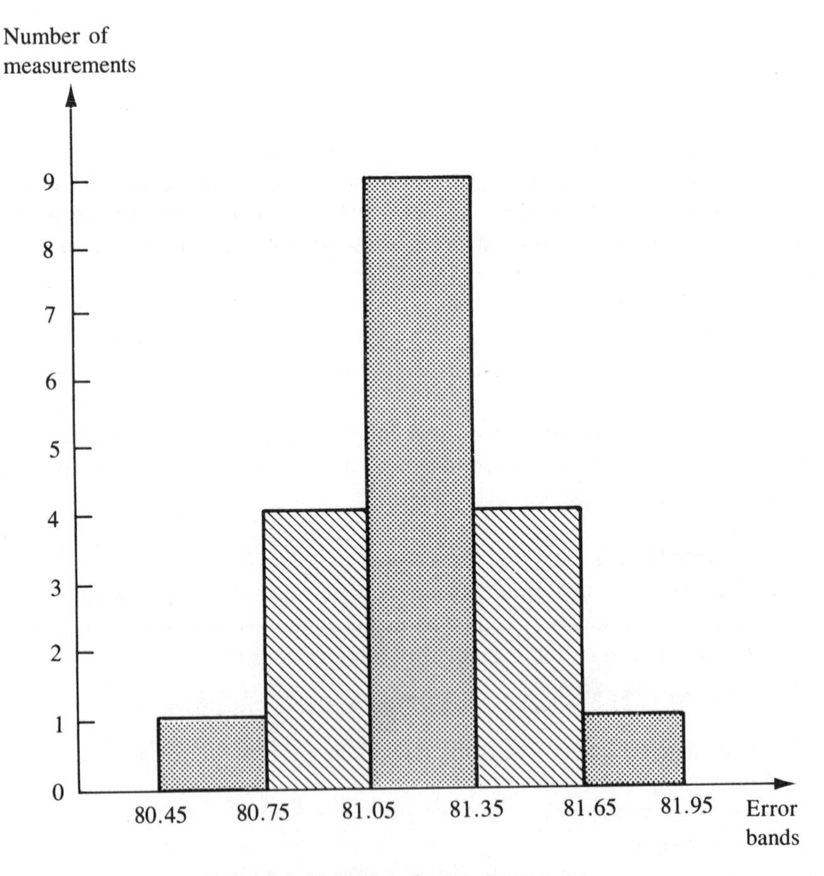

Figure 3.6 *Histogram of measurements*

3.3.2 Frequency distributions

A further and very powerful way of analyzing the pattern in which measurements deviate from the mean value is to use graphical techniques. The simplest way of doing this is by means of a *histogram*, where bands of equal width across the range of measurement values are defined and the number of measurements within each band are counted. Figure 3.6 shows a histogram of measurements drawn from the set of mass data in Table 3.1 by choosing bands 0.3 g wide. There are, for instance, nine measurements in the range between 81.05 g and 81.35 g, and so the height of the histogram at this point is nine units. (NB The scaling of the bands was deliberately chosen so that no measurements fell on the boundary between different bands and caused ambiguity about which band to put them in.) Such a histogram has the characteristic shape shown by truly random data, with symmetry about the mean value of the measurements.

As the number of measurements increases, smaller bands can be defined for the histogram, which retains its basic shape but then consists of a larger number of smaller steps on each side of the peak. In the limit, as the number of measurements approaches infinity, the histogram becomes a smooth curve known as the *frequency distribution curve* of the measurements, as shown in Figure 3.7. The ordinate of this curve is the frequency of occurrence of each measurement value, $F(X)$, and the abscissa is the

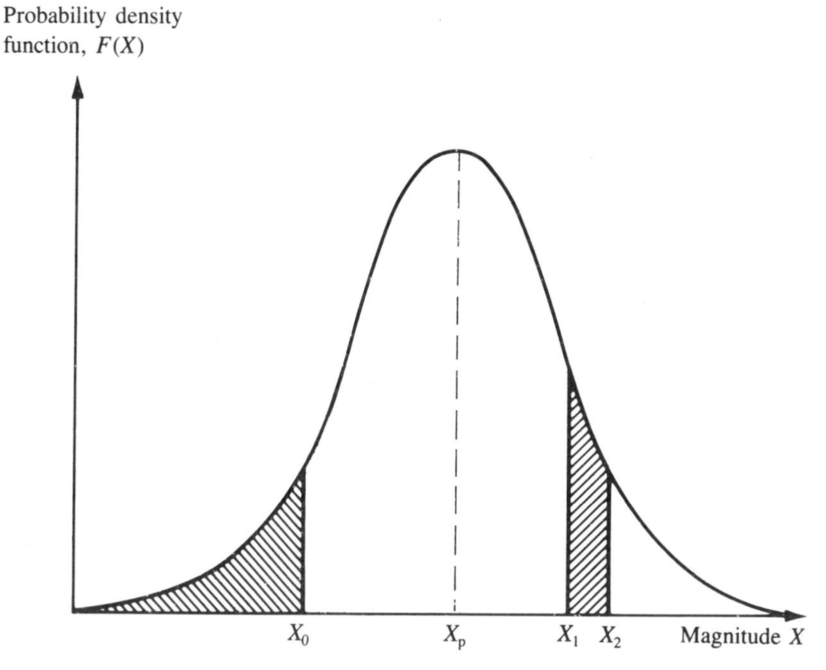

Figure 3.7 *Frequency distribution curve of measurements*

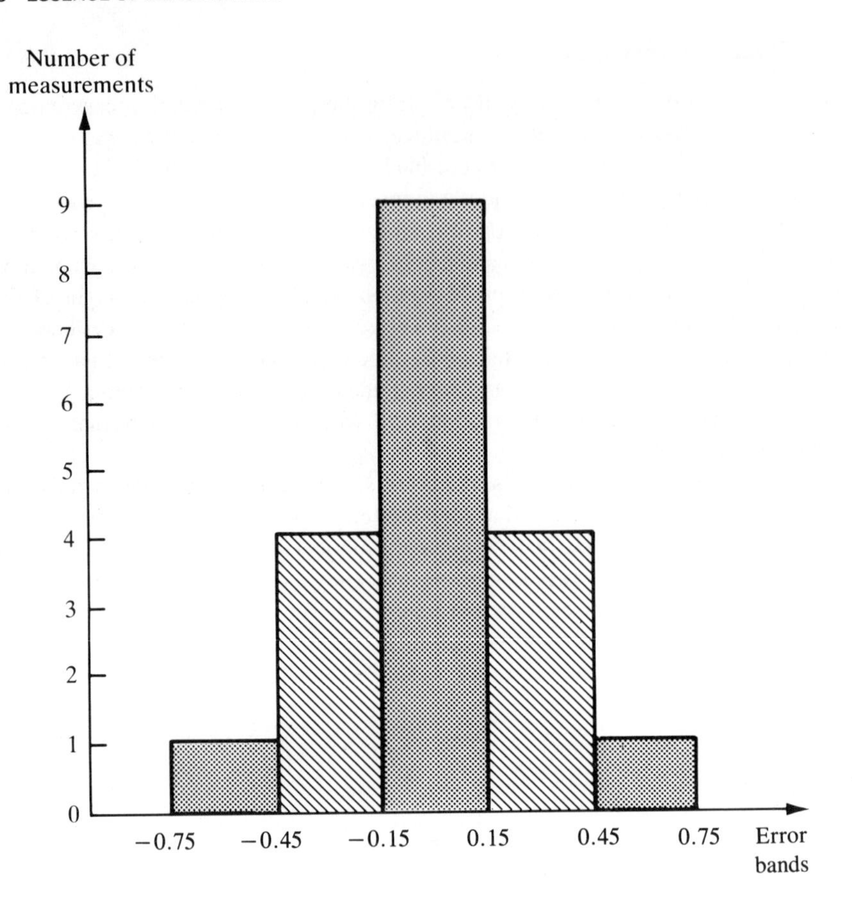

Figure 3.8 *Histogram of errors*

magnitude, X, and X_p is the most probable data value. If errors are truly random, X_p is the mean value of the measurements.

The symmetry of the measurements about the mean data value in Figures 3.6 and 3.7 is very useful for showing graphically that the measurement data have only random errors. However, in order to quantify the magnitude and distribution of the measurement errors, it is more useful to draw histograms and frequency distribution curves of the errors. To do this, the mean of the measurement data values is calculated first, then the error in each measurement in terms of its deviation from this mean value is calculated. Error bands of equal width are then defined and a histogram of errors drawn, as shown in Figure 3.8, according to the number of error values falling within each band. Provided that the errors are only random, this histogram has symmetry about the line of zero error. As the number of measurements increases, smaller error bands can be defined for the error histogram and in the limit, as the number of measurements approaches infinity, the histogram becomes a smooth curve as before.

In this case, the curve is known as a *frequency distribution curve* of the errors, as shown in Figure 3.9. The ordinate of this curve is the frequency of occurrence of each error level, $F(E)$, and the abscissa is the error magnitude, E. The error magnitude E_p corresponding to the peak of the frequency distribution curve is the value of error which has the greatest probability. If the errors are entirely random in nature, then the value of E_p will equal zero. Any non-zero value of E_p indicates systematic errors in the data, in the form of a bias which is often removable by recalibration.

If the height of the frequency distribution of errors curve is normalized such that the area under it is unity, then the curve in this form is known as a *probability curve*, and the height $F(E)$ at any particular error magnitude E is known as the *probability density function* (p.d.f.). The condition that the area under the curve is unity can be expressed mathematically as:

$$\int_{-\infty}^{\infty} F(E)\mathrm{d}E = 1$$

The probability that the error in any one particular measurement lies between two levels E_1 and E_2 can be calculated by measuring the area under the curve contained between two vertical lines drawn through E_1 and E_2, as shown by the right-hand hatched area in Figure 3.9. This can be expressed mathematically as:

$$P(E_1 \leq E \leq E_2) = \int_{E_1}^{E_2} F(E)\mathrm{d}E \quad (3.7)$$

Expression (3.7) is often known as the *error function*.

Of particular importance for assessing the maximum error likely in any one measurement is the *cumulative distribution function* (c.d.f.). This is defined as the probability of observing a value less than or equal to E_0, and is expressed mathematically as:

$$P(E \leq E_0) = \int_{-\infty}^{E_0} F(E)\mathrm{d}E \quad (3.8)$$

Thus the c.d.f. is the area under the curve to the left of a vertical line drawn through E_0, as shown by the left-hand hatched area on Figure 3.9.

Three special types of frequency distribution known as the Gaussian, binomial and Poisson distributions exist, and these are very important because most data sets approach closely to one or other of them. The distribution of relevance to data sets containing random measurement errors is the Gaussian one.

Gaussian distribution

A Gaussian curve is defined as a normalized frequency distribution where the frequency and magnitude of quantities are related by the expression:

$$F(E) = \frac{1}{\sqrt{\sigma(2\pi)}} \cdot \exp[-(x-m)^2/2\sigma^2] \quad (3.9)$$

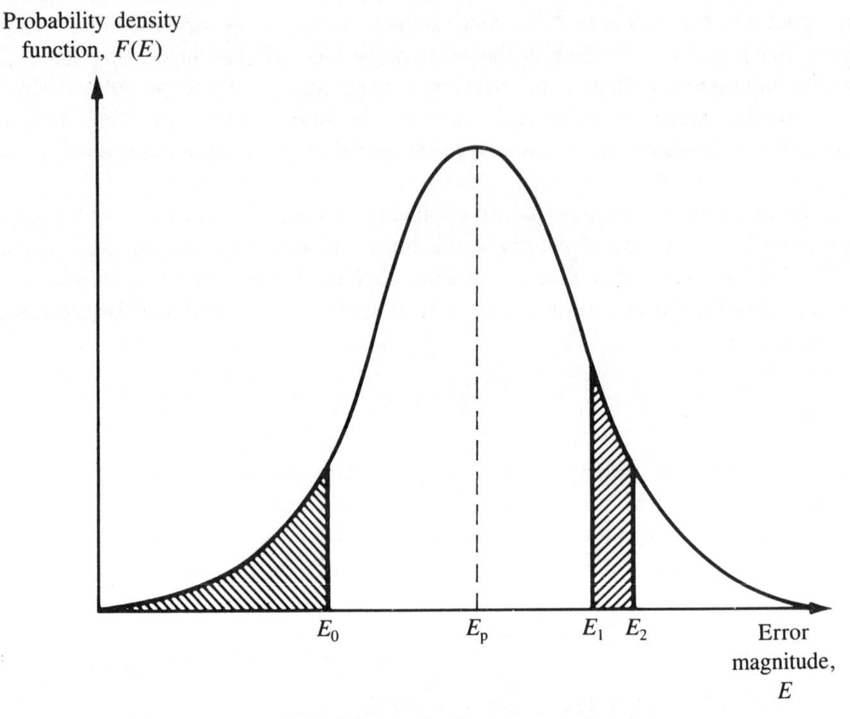

Figure 3.9 ***Frequency distribution curve of errors***

where m is the mean value of the measurement set x and the other quantities are as defined before. It is only applicable to data which has only random errors, i.e. where no systematic errors exist. Equation (3.9) is particularly useful for analyzing a Gaussian set of measurements and predicting how many measurements lie within some particular defined range.

If the measurement errors E are calculated for all measurements such that $E = x - m$, then the curve of error frequency $F(E)$ plotted against error magnitude E is a Gaussian curve known as the error frequency distribution curve. The mathematical relationship between $F(E)$ and E can then be derived by modifying equation (3.9) to give:

$$F(E) = \frac{1}{\sqrt{\sigma(2\pi)}} \cdot \exp[-E^2/2\sigma^2] \tag{3.10}$$

Most measurement data sets such as the values of mass in Table 3.1 fit to a Gaussian distribution curve because, if errors are truly random, small deviations from the mean value occur much more often than large deviations, i.e. the number of small errors is much larger than the number of large ones. Alternative names for the Gaussian distribution curve are the *normal distribution* or *bell-shaped distribution*.

The Gaussian distribution curve is symmetrical about the line through the mean of the measurement values, which means that positive errors away from the mean value occur in equal quantities to negative errors in any data set containing measurements subject to random error. If the standard deviation is used as a unit of error, the curve can be used to determine what probability there is that the error in any particular measurement in a data set is greater than a certain value. By substituting the expression for $F(E)$ (3.10) into the probability equation (3.7), the probability that the error lies in a band between error levels E_1 and E_2 can be expressed as:

$$P(E_1 \leq E \leq E_2) = \int_{E_1}^{E_2} \frac{1}{\sqrt{\sigma(2\pi)}} \cdot \exp[-E^2/2\sigma^2]\mathrm{d}E \qquad (3.11)$$

Equation (3.11) can be simplified by making the substitution

$$z = E/\sigma \qquad (3.12)$$

Then:

$$P(E_1 \leq E \leq E_2) = \int_{z_1}^{z_2} \frac{1}{\sqrt{\sigma(2\pi)}} \cdot \exp[-z^2/2]\mathrm{d}z \qquad (3.13)$$

Even after carrying out this simplification, equation (3.13) still cannot be evaluated by the use of standard integrals. Instead, numerical integration has to be used. To simplify the burden involved in this, standard error function tables have been drawn up which evaluate the integral for various values of z.

Error function tables

An error function table can be found in Appendix 1. This gives values of $F(z)$ for various values of z. $F(z)$ represents the proportion of data values which are less than or equal to z and is equal to the area under the normalized probability curve to the left of z. Study of the table will show that $F(z) = 0.5$ for $z = 0$. This shows that, as expected, the number of data values ≤ 0 is 50% of the total. This must be so if the data have only random errors.

Use of error function table:s

It will be observed that the table in Appendix 1, in common with most published error function tables, only gives $F(z)$ for positive values of z. For negative values of z, we can make use of the following relationship because the frequency distribution curve is normalized:

$$F(-z) = 1 - F(z) \qquad (3.14)$$

($F(-z)$ is the area under the curve to the left of $(-z)$, i.e. it represents the proportion of data vales $\leq -z$.)

EXAMPLE 3.3

How many measurements in a data set subject to random errors lie outside boundaries of $+\sigma$ and $-\sigma$, i.e. how many measurements have an error $< |\sigma|$?

SOLUTION

The required number is represented by the sum of the two shaded areas in Figure 3.10. This can be expressed mathematically as:

$$P[E < -\sigma \text{ or } E > +\sigma] = P[E < -\sigma] + P[E > +\sigma]$$

For $E = -\sigma$, $z = -1.0$ (from equation 3.12). Using an error function table:

$$P[E < -\sigma] = F(-1) = 1 - F(1) = 1 - 0.8413 = 0.1587$$

Similarly, for $E = +\sigma$, $z = +1.0$, the error function table gives:

$$P[E > +\sigma] = 1 - P[E < +\sigma] = 1 - F(1) = 1 - 0.8413 = 0.1587$$

(This last step is valid because the frequency distribution curve is normalized such that the total area under it is unity.) Thus

$$P[E < -\sigma] + P[E > +\sigma] = 0.1587 + 0.1587 = 0.3174 \approx 32\%$$

Therefore, 32% of the measurements lie outside the $\pm\sigma$ boundaries, i.e. 32% of the measurements have an error greater than $|\sigma|$. It follows that 68% of the measurements lie inside the boundaries of $\pm\sigma$.

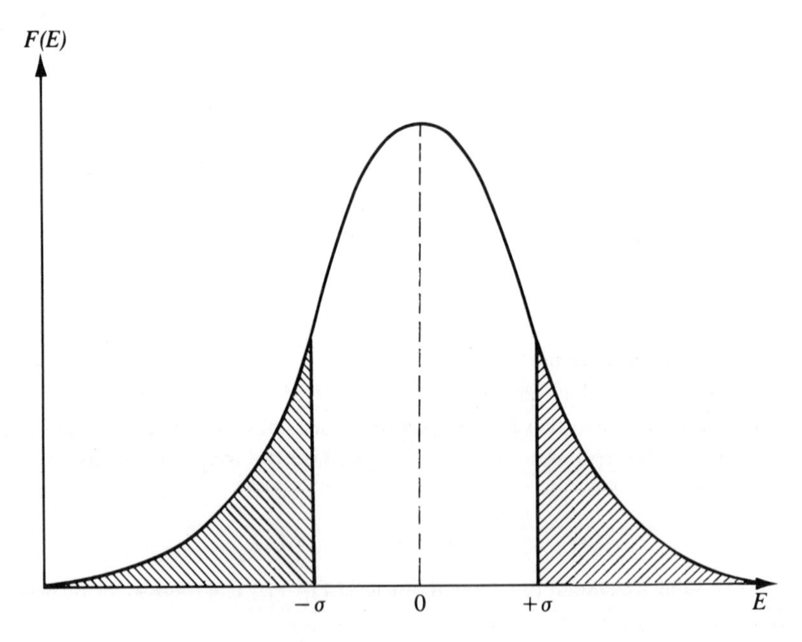

Figure 3.10 *$\pm\sigma$ boundaries on measurement errors*

Table 3.2

Error boundaries	*Percentage of data points within boundary*	*Probability of any particular data point being outside boundary (%)*
$\pm\sigma$	68.0	34.0
$\pm 2\sigma$	95.4	4.6
$\pm 3\sigma$	99.7	0.3

The above analysis shows that, for Gaussian-distributed data values, 68% of the measurements have errors which lie within the bounds of $\pm\sigma$. Similar analysis shows that boundaries of $\pm 2\sigma$ contain 95.4% of data points, and extending the boundaries to $\pm 3\sigma$ encompasses 99.7% of data points. The probability of any data point lying outside particular error boundaries can therefore be expressed by Table 3.2.

Distribution of manufacturing tolerances

The Gaussian distribution curve can be extended to analyze tolerances in manufactured components rather than errors in process measurements. In this form, it describes the frequency distribution of measurements as shown in Figure 3.11.

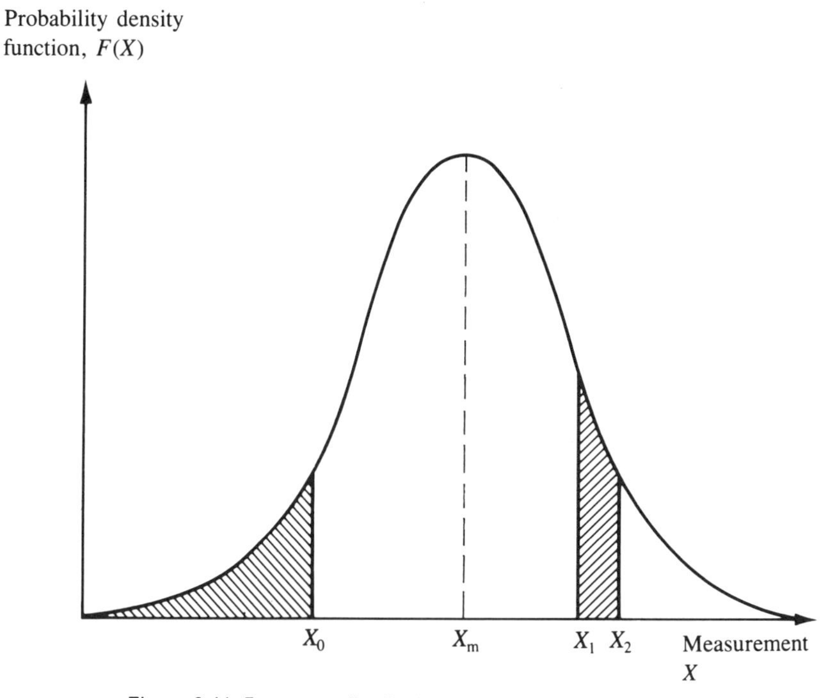

Figure 3.11 *Frequency distribution of manufacturing tolerances*

Here, $f(x)$ is the probability that the measurement has some particular value x. The most likely value of x, x_m, corresponds to the peak of the curve. If the measurements are Gaussian, the curve will be symmetrical about the line $x = x_m$, and x_m will represent the mean value of the measurements.

Equations similar to (3.7) and (3.8) can be written:

$$P(x_1 \le x \le x_2) = \int_{x_1}^{x_2} f(x)\mathrm{d}x \tag{3.15}$$

$$P(x \le x_0) = \int_{-\infty}^{x_0} f(x)\mathrm{d}x \tag{3.16}$$

Also, by modifying equation (3.13), we obtain:

$$P(x_1 \le x \le x_2) = \int_{x_1}^{x_2} \frac{1}{\sqrt{\sigma(2\pi)}} \cdot \exp[-(x-\mu)^2/2\sigma^2]\mathrm{d}x \tag{3.17}$$

where μ is the mean of the measurements ($= x_m$).

Having found the probability of any measurement chosen at random lying within the range of x_1 to x_2, the number of measurements N lying within the range of x_1 to x_2 can be calculated as follows:

$$N = \int_{x_1}^{x_2} \frac{n}{\sqrt{\sigma(2\pi)}} \cdot \exp[-(x-\mu)^2/2\sigma^2]\mathrm{d}x \tag{3.18}$$

where n is the total number of measurements.

If the substitution $z = (x-\mu)/\sigma$ is made, equation (3.18) simplifies to:

$$N = \int_{z_1}^{z_2} \frac{n}{\sqrt{\sigma(2\pi)}} \cdot \exp[-z^2/2]\mathrm{d}z \tag{3.19}$$

This is now in a form which can be evaluated using an error function table (see Appendix 1).

EXAMPLE 3.4

An integrated circuit chip contains 10^5 transistors. The transistors have a mean current gain of 20 and a standard deviation of 2. Calculate (a) The number of transistors with a current gain between 19.8 and 20.2, and (b) The number of transistors with a current gain greater than 17.

SOLUTION

(a) The proportion of transistors where $19.8 < \text{gain} < 20.2$ is given by $P[x < 20.2] - P[x < 19.8]$. Calculating z from $z = (x-\mu)/\sigma$: for $x = 20.2$; $z = 0.1$ and for $x = 19.8$; $z = -0.1$. From the error function table, $P[z < 0.1] = 0.5398$

$$P[z < -0.1] = 1 - P[z < 0.1] = 1 - 0.5398 = 0.4602$$

Thus,

$$P[x < 20.2] - P[x < 19.8] = P[z < 0.1] - P[z < -0.1]$$
$$= 0.5398 - 0.4602 = 0.0796$$

Thus $0.0796 \times 10^5 = 7960$ transistors have a current gain in the range from 19.8 to 20.2.

(b) The number of transistors with gain > 17 is given by:

$$P[x > 17] = 1 - P[x < 17] = 1 - P[z < -1.5] = P[z < +1.5] = 0.9332$$

Thus, 93.32%, i.e. 93 320 transistors, have a gain > 17.

Standard error of the mean

The foregoing analysis is only strictly true for measurement sets containing infinite populations. It is not of course possible to obtain an infinite number of data values, and some error must therefore be expected in the calculated mean value of the practical, finite data set available. If several subsets are taken from an infinite data population, then, by the central limit theorem, the means of the subsets will form a Gaussian distribution about the mean of the infinite data set. The error in the mean of a finite data set is usually expressed as the *standard error of the mean*, α, which is calculated as:

$$\alpha = \sigma / \sqrt{n}$$

This tends towards zero as the number of measurements in the data set is expanded towards infinity. The value obtained from a set of n measurements $x_1, x_2, \ldots, x_n$ is then expressed as:

$$x = x_{\text{mean}} \pm \alpha$$

For the data set of mass measurements in Table 3.1, $n = 19$, $\sigma = 0.318$ and $\alpha = 0.073$. The mass can therefore be expressed as 81.18 ± 0.07 kg (68% confidence limit). However, it is more usual to express measurements with 95% confidence limits ($\pm 2\sigma$ boundaries). In this case, $2\sigma = 0.636$, $2\alpha = 0.146$ and the value of the mass can therefore be expressed as 81.18 ± 0.15 kg (95% confidence limit).

3.4 Error reduction using intelligent instruments

Large reductions in measurement errors can often be achieved by using intelligent instruments. An intelligent instrument comprises all the usual elements of a measurement system as shown in Figure 1.2. However, it is distinguished from dumb (non-intelligent) measurement systems by the inclusion of a microcomputer and by the addition of one or more extra transducers at its input.

The inputs to an intelligent instrument are data from a *primary transducer* and additional data from one or more *secondary transducers*. The primary transducer measures the magnitude of the main quantity of interest while the secondary transducers measure the magnitude of environmental parameters. For instance, in an intelligent mass measuring instrument, the primary transducer is usually a load cell and additional secondary transducers are provided to measure modifying inputs such as the ambient temperature and atmospheric pressure.

The microcomputer performs preprogrammed signal-processing functions and data manipulation algorithms on the data from the primary transducer (the measured quantity of interest), using data read from the secondary transducer(s), and outputs the processed measurement from the primary transducer for presentation at the instrument output. The effect of this computerization of the signal-processing function is an improvement in the quality of the instrument's output measurements and a general simplification of the signal-processing task. Some examples of the signal processing which a microprocessor within an intelligent instrument can readily perform include correction of the instrument output for bias caused by environmental variations (e.g. temperature changes), and conversion to produce a linear output from a transducer whose characteristic is fundamentally non-linear. A fuller discussion about the techniques of digital signal processing can be found in Section 5.7.

An intelligent instrument behaves as a black box as far as the user is concerned, and no knowledge of its internal mode of operation is required in normal measurement situations. However, the following summary of its operating principles may be of interest to some readers.

Like computer systems in general, the computer hardware within an intelligent instrument contains the essential components of a central processing unit, some memory and an input–output interface. Both ROM (read only memory) and RAM (random access memory) types of memory are included, ROM to store the signal-processing program and RAM to provide space to store input–output data. As the external data signals are usually analog in form, it is normally necessary to include analog-to-digital and digital-to-analog conversion elements within the input–output interface.

The signal-processing operation requires data values to be input, processed and output according to a sequence of operations defined by the computer program. It is not usual for the user to be expected to write this program. Indeed, there is rarely any provision for the user to create or modify operating programs even if he or she wished to do so. There are several reasons for this. Firstly, the signal processing needed within an intelligent instrument is usually well defined, and therefore it is more efficient for a manufacturer to produce this rather than to have each individual user produce near identical programs separately. Secondly, better program integrity and instrument operation are achieved if a standard program produced by the instrument manufacturer is used. Finally, use of a standard program allows it to be burnt into ROM, thereby protecting it from any failure of the instrument power supply. This also facilitates software maintenance and updates, by the mechanism of the manufacturer providing a new ROM which simply plugs into the slot previously occupied by the old ROM.

Intelligent instruments offer many advantages over their non-intelligent counterparts, principally because of the improvement in accuracy achieved by processing the output of transducers to correct for errors inherent in the measurement process. The solutions which intelligent instruments offer to many problems occurring in measurement systems will be discussed at various points in the chapters following.

One example of the benefit that intelligence can bring to instruments is in volume flow rate measurement, where the flow rate is inferred by measuring the differential pressure across an orifice plate placed in a fluid-carrying pipe (see Chapter 9 for more details). The flow rate is proportional to the square root of the difference in pressure across the orifice plate. For a given flow rate, this relationship is affected both by the temperature and by the mean pressure in the pipe, and changes in the ambient value of either of these cause measurement errors. A typical intelligent flow rate measuring instrument contains three transducers, a primary one measuring the pressure difference across an orifice plate and secondary ones measuring absolute pressure and temperature. The instrument is programmed to correct the output of the primary differential pressure transducer according to the values measured by the secondary transducers, using appropriate physical laws which quantify the effect of ambient temperature and pressure changes on the fundamental relationship between flow and differential pressure. The instrument is also normally programmed to convert the square root relationship between flow and signal output into a direct one, making the output much easier to interpret. Typical inaccuracy levels of such intelligent flow measuring instruments are ±0.1%, compared with ±0.5% for their non-intelligent equivalents, showing an improvement by a factor of five.

Intelligent instruments usually provide many other facilities in addition to those mentioned above, such as the following:

1 Signal damping with selectable time constants.
2 Switchable ranges (using several primary transducers within the instrument which each measure over a different range).
3 Switchable output units (e.g. display in imperial or SI units).
4 Diagnostic facilities.
5 Remote adjustment and control of instrument options from up to 1500 metres away via four-way, 20 mA signal lines.

Suitable care must always be taken when introducing a microcomputer into a measurement system to avoid creating new sources of measurement noise. This is particularly so where one microcomputer is used to process the output of several transducers and is connected to them by signal wires. In such circumstances, the connections and connecting wires can create noise through electrochemical potentials, thermoelectric potentials, offset voltages introduced by common mode impedances, and a.c. noise at power, audio and radio frequencies. Recognition of all these possible noise sources allows them to be eliminated in most cases by employing good measurement system construction practices. All remaining noise sources are usually eliminated by the provision of a set of four earthing circuits within the interface which fulfil the following functions:

1 Power earth: provides a path for fault currents due to power faults.
2 Logic earth: provides a common line for all logic circuit potentials.

3 Analog earth (ground): provides a common reference for all analog signals.
4 Safety earth: connected to all metal parts of equipment to protect personnel should power lines come into contact with metal enclosures.

3.4.1 Reduction of systematic errors

The inclusion of intelligence in instruments can bring about a gross reduction in the magnitude of systematic errors. For instance, in the case of electrical circuits which are disturbed by the loading effect of the measuring instrument, an intelligent instrument can readily correct for measurement errors by applying equations such as (3.3) with the resistance of the measuring instrument inserted.

Intelligent instruments are particularly effective in improving the accuracy of measurements subject to modifying inputs, by the mechanism of the computer within the instrument correcting the measurement obtained from the primary transducer according to the values read by the secondary transducers. However, their ability to achieve this requires that the following preconditions be satisfied:

1 The physical mechanism by which a measurement transducer is affected by ambient condition changes must be fully understood and all physical quantities which affect the transducer output must be identified.
2 The effect of each ambient variable on the output characteristic of the measurement transducer must be quantified.
3 Suitable secondary transducers for monitoring the value of all relevant ambient variables must be available for input to the intelligent instrument.

Condition (1) means that the thermal expansion/contraction of all elements within a transducer must be considered in order to evaluate how it will respond to ambient temperature changes. Similarly, the transducer response, if any, to changes in ambient pressure, humidity, gravitational force must be examined.

Quantification of the effect of each ambient variable on the characteristics of the measurement transducer is then necessary, as stated in condition (2). Analytic quantification of ambient condition changes from purely theoretical consideration of the construction of a transducer is usually extremely complex and so is normally avoided. Instead, the effect is quantified empirically in laboratory tests where the output characteristic of the transducer is observed as the ambient environmental conditions are changed in a controlled manner.

Once the ambient variables affecting a measurement transducer have been identified and their effect quantified, an intelligent instrument can be designed which includes secondary transducers to monitor the value of the ambient variables. Suitable transducers which will operate satisfactorily within the environmental conditions prevailing for the measurement situation must of course exist, as stated in condition (3).

3.4.2 Reduction of random errors

If a measurement system is subject to random errors, intelligent instruments can be programmed to take a succession of measurements of a quantity within a short space of time and perform simple averaging or other statistical techniques on the readings before displaying an output measurement. This is valid for reducing any form of random error, including those due to human observation deficiencies, electrical noise or other random fluctuations.

As well as displaying an average value obtained from a number of measurements, intelligent instruments are often able to display other statistical parameters about the measurements taken, such as the standard deviation, variance and standard error of the mean. All of these quantities could of course be calculated manually, but the great advantage of using intelligent instruments is their much higher processing speed and the avoidance of the arithmetic errors which are liable to occur if humans perform these functions.

3.5 Total measurement system errors

A measurement system often consists of several separate components, each of which is subject to systematic and/or random errors. Mechanisms have now been presented for quantifying the errors arising from each of these sources and therefore the total error at the output of each measurement system component can be calculated. What remains to be investigated is how the errors associated with each measurement system component combine together, so that a total error calculation can be made for the complete measurement system.

All four mathematical operations of addition, subtraction, multiplication and division may be performed on measurements derived from different instruments/ transducers in a measurement system. Appropriate techniques for the various situations which arise are covered below.

3.5.1 *Error in a product*

If the outputs y and z of two measurement system components are multiplied together, the product can be written as:

$$P = yz$$

If the possible error in y is $\pm ay$ and in z is $\pm bz$, then the maximum and minimum values possible in P can be written as:

$$P_{max} = (y + ay)(z + bz) \qquad P_{min} = (y - ay)(z - bz)$$
$$= yz + ayz + byz + aybz \qquad = yz - ayz - byz + aybz$$

For typical measurement system components with output errors of up to 1% or 2% in magnitude, both a and b are very much less than one in magnitude and thus terms in $aybz$ are negligible compared with other terms. Therefore we have:

$$P_{max} = yz(1 + a + b) \quad P_{min} = yz(1 - a - b)$$

Thus the possible error in the product P lies within the range of $\pm(a + b)$.

EXAMPLE 3.5

If the power in a circuit is calculated from measurements of voltage and current in which the calculated maximum errors are respectively ±1% and ±2%, then the possible error in the calculated power value is ±3%.

3.5.2 *Error in a quotient*

If the output measurement y of one system component with possible error $\pm ay$ is divided by the output measurement z of another system component with possible error $\pm bz$, then the maximum and minimum possible values for the quotient can be written as:

$$Q_{max} = \frac{y + ay}{z - bz} = \frac{(y + ay)(z + bz)}{(z - bz)(z + bz)} = \frac{yz + ayz + byz + abyz}{z^2 - b^2z^2}$$

$$Q_{min} = \frac{y - ay}{z + bz} = \frac{(y - ay)(z - bz)}{(z + bz)(z - bz)} = \frac{yz - ayz - byz + abyz}{z^2 - b^2z^2}$$

For $a \ll 1$ and $b \ll 1$, terms in ab and b^2 are negligible compared with the other terms. Hence:

$$Q_{max} \approx \frac{yz(1 + a + b)}{z^2} \qquad Q_{min} \approx \frac{yz(1 - a - b)}{z^2}$$

i.e.

$$Q = \frac{y}{z} \pm \frac{y}{z}(a + b)$$

Thus the possible error in the quotient lies within the range $\pm(a + b)$.

EXAMPLE 3.6

If the resistance in a circuit is calculated from measurements of voltage and current where the respective errors are ±1% and ±0.3%, the likely error in the resistance value is ±1.3%.

3.5.3 *Error in a sum*

If the two outputs y and z of separate measurement system components are to be added together, we can write the sum as:

$$S = y + z$$

If the maximum errors in y and z are $\pm ay$ and $\pm bz$ respectively, we can express the maximum and minimum possible values of S as:

$$S_{max} = y + ay + z + bz \qquad S_{min} = y - ay + z - bz$$

or:

$$S = y + z \pm (ay + bz)$$

This relationship for S is not convenient because in this form the error term cannot be expressed as a fraction or percentage of the calculated value for S. Fortunately, statistical analysis can be applied which expresses S in an alternative form such that the most probable maximum error in S is represented by a quantity e, where e is given by:

$$e = \sqrt{(ay)^2 + (bz)^2} \tag{3.20}$$

Thus:

$$S = (y + z) \pm e$$

This can be expressed in the alternative form:

$$S = (y + z)(1 \pm f) \tag{3.21}$$

where $f = e/(y + z)$.

EXAMPLE 3.7

A circuit requirement for a resistance of 550 Ω is satisfied by connecting together two resistors of nominal values 220 Ω and 330 Ω in series. If each resistor has a tolerence of ±2%, the error in the sum calculated according to equations (3.20) and (3.21) is given by:

$$e = \sqrt{(0.02 \times 220)^2 + (0.02 \times 330)^2} = 7.93$$
$$f = 7.93/550 = 0.0144$$

Thus the total resistance S can be expressed as:

$$S = 550\,\Omega \pm 7.93\,\Omega$$

or

$$S = 550(1 \pm 0.0144)\,\Omega,$$

i.e.

$$S = 550\,\Omega \pm 1.4\%.$$

3.5.4 Error in a difference

If the two outputs y and z of separate measurement systems are to be subtracted from one another, and the possible errors are $\pm ay$ and $\pm bz$, then the difference S can be expressed as:

$$S = (y - z) \pm e \quad \text{or} \quad S = (y - z)(1 \pm f)$$

where e is calculated as above (equation 3.20), and $f = e/(y - z)$.

EXAMPLE 3.8

A fluid flow rate is calculated from the difference in pressure measured on both sides of an orifice plate. If the pressure measurements are 10.0 and 9.5 bar and the error in the pressure measuring instruments is specified as ±0.1%, then values for e and f can be calculated as:

$$e = \sqrt{(0.001 \times 10)^2 + (0.001 \times 9.5)^2} = 0.0138$$

$$f = 0.0138/0.5 = 0.0276$$

Thus the pressure difference can be expressed as 0.5 bar ± 2.8%.

This example illustrates very poignantly the relatively large error which can arise when calculations are made based on the difference between two measurements.

3.5.5 Total error when combining multiple measurements

The final case to be covered is where the final measurement is calculated from several measurements which are combined together in a way which involves more than one type of arithmetic operation. For example, the density of a rectangular-sided solid block of material can be calculated from measurements of its mass divided by the product of measurements of its length, height and width. The errors involved in each stage of arithmetic are cumulative, and so the total measurement error can be calculated by adding together the two error values associated with the two multiplication stages involved in calculating the volume and then calculating the error in the final arithmetic operation when the mass is divided by the volume.

EXAMPLE 3.9

A rectangular-sided block has edges of lengths a, b and c, and its mass is m. If the values and possible errors in quantities a, b, c and m are as shown below, calculate the value of density and the possible error in this value: a = 100 mm ±1%, b = 200 mm ±1%, c = 300 mm ±1%, m = 20 kg ±0.5%.

SOLUTION

Value of ab = 0.02 m^2 ±2% (possible error = 1% + 1% = 2%)
Value of $(ab)c$ = 0.006 m^3 ±3% (possible error = 2% + 1% = 3%)
Value of $m/(abc)$ = 20/0.006 = 3330 kg/m^3 ±3.5% (possible error = 3% + 0.5% = 3.5%)

3.6 Self-assessment questions

3.1 Explain the difference between systematic and random errors. What are the typical sources of these two types of error?

3.2 In what ways can the act of measurement cause a disturbance in the system being measured?

3.3 Suppose that the components in the circuit shown in Figure 3.1 have the following values: $R_1 = 1\,\text{k}\Omega$; $R_2 = 2\,\text{k}\Omega$. If the instrument measuring the output voltage across AB has a resistance of $10\,\text{k}\Omega$, what is the measurement error caused by the loading effect of this instrument?

3.4 Instruments are normally calibrated and their characteristics defined for particular standard ambient conditions. What procedures are normally taken to avoid measurement errors when using instruments which are subjected to changing ambient conditions?

3.5 What steps can be taken to reduce the effect of modifying inputs in measurement systems?

3.6 The output of a potentiometer is measured by a voltmeter having a resistance R_m, as shown in Figure 3.12. R_t is the resistance of the total length X_t of the potentiometer and R_i is the resistance between the wiper and common point C for a general wiper position X_i. Show that the measurement error due to the resistance R_m of the measuring instrument is given by:

$$\text{Error} = E\frac{R_i^2(R_t - R_i)}{R_t(R_i R_t + R_m R_t - R_i^2)}$$

Hence show that the maximum error occurs when X_i is approximately equal to $2X_t/3$.

(Hint: differentiate the error expression with respect to R_i and set equal to 0. Note that maximum error does not occur exactly at $X_i = 2X_t/3$, but this value is very close to the position where the maximum error occurs.)

3.7

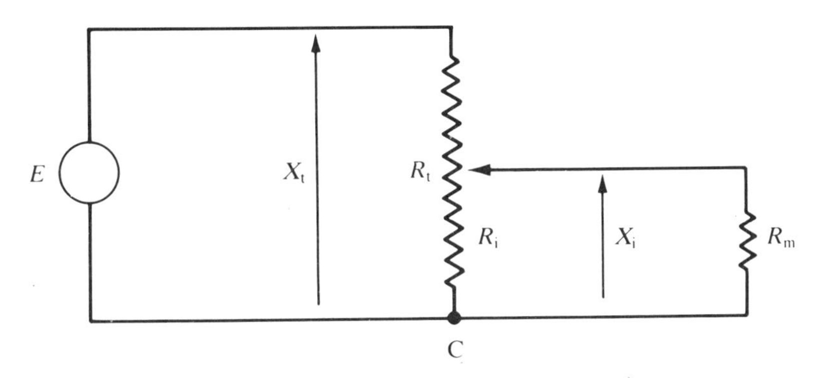

Figure 3.12 *Circuit for self-assessment question 3.6*

In a survey of fifteen owners of a certain model of car, the following figures for average petrol consumption in miles/gallon were reported:

25.5 30.3 31.1 29.6 32.4 39.4 28.9 30.0
33.3 31.4 29.5 30.5 31.7 33.0 29.2

Calculate the mean value, the median value and the standard deviation of the data set.

3.8 (a) Explain the term probability density function.
(b) Write down an expression for a Gaussian probability density function of given mean value μ and standard deviation σ and show how you would obtain the best estimate of these two quantities from a sample of population n.
(c) The following ten measurements are made of the output voltage from a high-gain amplifier which is contaminated due to noise fluctuations:

1.53, 1.57, 1.54, 1.54, 1.50, 1.51, 1.55, 1.54, 1.56, 1,53

Determine the mean value and standard deviation. Hence estimate the accuracy to which the mean value is determined from these ten measurements. If one thousand measurements were taken, instead of ten, but σ remained the same, by how much would the accuracy of the calculated mean value be improved?

3.9 The measurements in a data set are subject to random errors but it is known that the data set fits a Gaussian distribution. Use an error function table to determine the percentage of measurements which lie within the boundaries of $\pm 1.5\sigma$, where σ is the standard deviation of the measurements.

3.10 The thickness of a set of gaskets varies because of random manufacturing disturbances but the thickness values measured belong to a Gaussian distribution. If the mean thickness is 3 mm and the standard deviation is 0.25, calculate the percentage of gaskets which have a thickness greater than 2.5 mm.

3.11 A 3 volt d.c. power source required for a circuit is obtained by connecting together two 1.5 V batteries in series. If the error in the voltage output of each battery is specified as ±1%, calculate the possible error in the 3 volt power source which they make up.

3.12 In order to calculate the heat loss through the wall of a building, it is necessary to know the temperature difference between the inside and outside walls. If temperatures of 5 and 20 °C are measured on each side of the wall by mercury-in-glass thermometers with a range of −25 to +25 °C and a quoted inaccuracy figure of ±1% of full-scale reading, calculate the possible error in the calculated figure for the temperature difference.

3.13 The power dissipated in a car headlight is calculated by measuring the direct current (d.c.) voltage drop across it and the current through it ($P = V \times I$). If the possible errors in the measured voltage and current values are ±1% and ±2% respectively, calculate the possible error in the power value deduced.

3.14 The resistance of a carbon resistor is measured by applying a d.c. voltage across it and measuring the current ($R = V/I$). If the voltage and current values are measured as 10±0.1 V and 214±5 mA respectively, calculate the value of the carbon resistor.

3.15 The density (d) of a liquid is calculated by measuring its depth (c) in a calibrated rectangular tank and then emptying it into a mass-measuring system. The length and width of the tank are (a) and (b) respectively and thus the density is given by:

$$d = m/(a \times b \times c)$$

where m is the measured mass of the liquid emptied out. If the possible errors in the measurements of a, b, c and m are 1%, 1%, 2% and 0.5% respectively, determine the possible error in the calculated value of the density (d).

3.16 The volume flow rate of a liquid is calculated by allowing the liquid to flow into a cylindrical tank (stood on its flat end) and measuring the height of the liquid surface before and after the liquid has flowed for 10 minutes. The volume collected after 10 minutes is given by:

$$\text{volume} = (h_2 - h_1)\pi(d/2)^2$$

where h_1 and h_2 are the starting and finishing surface heights and d is the measured diameter of the tank.

(a) If $h_1 = 2$ m, $h_2 = 3$ m and $d = 2$ m, calculate the volume flow rate in m^3/min.

(b) If the possible error in each measurement h_1, h_2 and d is $\pm 1\%$, estimate the possible error in the calculated value of volume flow rate.

References and further reading

Morris, A.S., *Principles of Measurement and Instrumentation*, 1993. Prentice Hall, Hemel Hempstead.

Chatfield, C., *Statistics for Technology*, 1983. Chapman and Hall, London.

CHAPTER 4

Instrument calibration

4.1 Introduction

Instrument calibration consists of comparing the output of the instrument under test against the output of an instrument of known accuracy when the same input is applied to both instruments. This procedure is carried out for a range of inputs covering the whole measurement range of the instrument. Calibration ensures that the measuring accuracy of all instruments used in a measurement system is known over the whole measurement range, provided that the instruments are used in environmental conditions that are the same as those under which they were calibrated. For use of instruments under different environmental conditions, appropriate correction has to be made for the ensuing modifying inputs, as described in Chapter 3.

Instrument calibration has to be repeated at prescribed intervals because the characteristics of any instrument change over a period of time. Changes in instrument characteristics are brought about by such factors as mechanical wear, and the effects of dirt, dust, fumes and chemicals in the operating environment. To a great extent, the magnitude of the drift in characteristics depends on the amount of use an instrument receives and hence on the amount of wear and the length of time that it is subjected to the operating environment. However, some drift even occurs when the instrument is in storage, as a result of ageing effects in components within the instrument.

The reasons for this drift in characteristics were discussed in detail in Chapter 3. It is sufficient here to accept that such drift does occur and that the rate at which characteristics change with time varies according to the type of instrument used, the frequency of use and the prevailing environmental conditions.

Because the rate of change of instrument characteristics is influenced by so many factors, it is difficult or, in some cases, impossible to determine the required frequency of instrument recalibration from theoretical considerations. Instead, practical experimentation is applied to determine the rate of such changes. Once the maximum permissible measurement error has been defined, knowledge of the rate at which the characteristics of an instrument change allows calculation of the time when an instrument will have reached the bounds of its acceptable performance level. The instrument must be recalibrated either at this time or earlier. This measurement error level which an instrument reaches just before recalibration is the error bound which must be quoted in the documented specifications for the instrument.

4.2 **Process instrument calibration**

Calibration consists of comparing the output of the process instrument being calibrated against the output of a standard instrument of known accuracy, when the same input (measured quantity) is applied to both instruments. During this calibration process, the instrument is tested over its whole range by repeating the comparison procedure for a range of inputs.

The instrument used as a standard for this procedure must be one which is kept solely for calibration duties. It must never be used for other purposes. Most particularly, it must not be regarded as a spare instrument which can be used for process measurements if the instrument normally used for that purpose breaks down. Proper provision for process instrument failures must be made by keeping a spare set of process instruments. Standard calibration instruments must be totally separate.

To ensure that these conditions are met, the calibration function must be managed and executed in a professional manner. This will normally mean setting aside a particular place within the instrumentation department of a company where all calibration operations take place and where all instruments used for calibration are kept. As far as possible this should take the form of a separate room, rather than a sectioned-off area in a room used for other purposes as well. This will enable better environmental control to be applied in the calibration area and will also offer better protection against unauthorized handling or use of the calibration instruments. The level of environmental control required during calibration should be considered carefully with due regard to the level of accuracy required in the calibration procedure, but should not be overspecified as this will lead to unnecessary expense. Full air conditioning is not normally required for calibration at this level, as it is very expensive, but sensible precautions should be taken to guard the area from extremes of heat or cold, and also good standards of cleanliness should be maintained. Useful guidance on the operation of standards facilities can be found elsewhere (British Standards Society, 1979).

While it is desirable that all calibration functions are performed in this carefully controlled environment, it is not always practical to achieve this. Sometimes, it is not convenient or possible to remove instruments from process plant, and in these cases it is standard practice to calibrate them *in situ*. In these circumstances, appropriate corrections must be made for the deviation in the calibration environmental conditions away from those specified. This practice does not obviate the need to protect calibration instruments and maintain them in constant conditions in a calibration laboratory at all times other than when they are involved in such calibration duties on plant.

Apart from the precautions taken to preserve the accuracy of instruments used for calibration, by treating them carefully and reserving them only for calibration duties, they are often chosen to be of a greater inherent accuracy than the process instruments that they are used to calibrate. Where instruments are only used for calibration purposes, greater accuracy can often be achieved by specifying a type of instrument which would be unsuitable for normal process measurements. Ruggedness, for instance, is not a requirement, and freedom from this constraint opens up a much wider range of possible instruments. In practice, high-accuracy, null-type instruments are very commonly used for calibration duties, because their requirement for a human operator is not a problem in these circumstances.

As far as management of calibration procedures is concerned, it is important that the performance of all calibration operations is assigned as the clear responsibility of just one person. That person should have total control over the calibration function, and be able to limit access to the calibration laboratory to designated approved personnel only. Only by giving this appointed person total control over the calibration function can the function be expected to operate efficiently and effectively. Lack of such definite management can only lead to unintentional neglect of the calibration system, resulting in the use of equipment in an out-of-date state of calibration and subsequent loss of traceability to reference standards. Professional management is essential so that the customer can be assured that an efficient calibration system is in operation and that the accuracy of measurements is guaranteed.

Calibration procedures which relate in any way to measurements which are used for quality control functions are controlled by the international standard BS.EN.ISO 9000 (this subsumes the old British quality standard BS 5750). One of the clauses in BS.EN.ISO 9000 requires that all persons using calibration equipment be adequately trained. The manager in charge of the calibration function is clearly responsible for ensuring that this condition is met. Training must be adequate and targeted at the particular needs of the calibration systems involved. People must understand what they need to know and especially why they must have this information. Successful completion of training courses should be marked by the award of qualification certificates. These attest to the proficiency of personnel involved in calibration duties and are a convenient way of demonstrating that the BS.EN.ISO 9000 training requirement has been satisfied.

The calibration facilities provided within the instrumentation department of a company provide the first link in the calibration chain. Instruments used for calibration at this level are known as working standards. A fundamental responsibility in the supervision of this calibration function is to establish the frequency at which the various shop-floor instruments should be calibrated and ensure that calibration is carried out at the appropriate times.

Determination of the frequency at which instruments should be calibrated is dependent upon several factors which require specialist knowledge. If an instrument is required to measure some quantity with a maximum inaccuracy of ±2%, then a certain amount of performance degradation can be allowed if its inaccuracy immediately after recalibration is ±1%. What is important is that the pattern of performance degradation be quantified, such that the instrument can be recalibrated before its inaccuracy has increased to the limit defined by the application.

The quantities which cause the deterioration in the performance of instruments over a period of time are mechanical wear, dust, dirt, ambient temperature and frequency of usage. Susceptibility to these factors varies according to the type of instrument involved. The effect of these quantities on the accuracy and other characteristics of an instrument can only be quantified by possessing an in-depth knowledge of the mechanical construction and other features involved in the instrument. Some form of practical experimentation is normally required to determine the necessary calibration frequency in the typical operating conditions for the instrument. Further discussion on the means of quantifying the rate of change of instrument characteristics was given in Chapter 3.

A proper course of action must be defined which describes the procedures to be followed when an instrument is found to be out of calibration, i.e. when its output is different from that of the calibration instrument when the same input is applied. The required action depends very much upon the nature of the discrepancy and the type of instrument involved. In many cases, deviations in the form of a simple output bias can be corrected by a small adjustment to the instrument, following which the adjustment screws must be sealed to prevent tampering. In other cases, the output scale of the instrument may have to be redrawn, or scaling factors altered where the instrument output is part of some automatic control or inspection system. In extreme cases, where the calibration procedure reveals signs of instrument damage, it may be necessary to send the instrument for repair or even scrap it.

Whatever system and frequency of calibration is established, it is important to review this from time to time to ensure that the system remains effective and efficient. It may happen that a cheaper but equally effective method of calibration becomes available with the passage of time, and such an alternative system must clearly be adopted in the interests of cost-efficiency. However, the main item under scrutiny in this review is normally whether the calibration interval is still appropriate. Records of the calibration history of the instrument will be the primary basis on which this review is made. It may happen that an instrument starts to go out of calibration more quickly after a period of time, either because of ageing factors within the instrument or because of changes in the operating environment. The conditions or mode of usage of the instrument may also be subject to change. As the environmental and usage conditions of an instrument may change beneficially as well as adversely, there is the possibility that the recommended calibration interval may increase as well as decrease.

Maintaining proper records is an important part of fulfilling this calibration function. A separate record, similar to that shown in Figure 4.1, should be kept for every instrument in the factory, whether it is in use or kept as a spare. This record should start by giving a description of the instrument and be followed by stating the required calibration frequency. Each occasion when the instrument is calibrated should be recorded in this record, and every such calibration log should show the status of the instrument in terms of the deviation from its required specification and the action taken to correct it. The calibration record is also very useful in providing feedback which shows whether the calibration frequency has been chosen correctly or not.

4.3 Standards laboratories

We have established so far that process instruments which are used to make quality-

Type of instrument		Company serial number:
Manufacturer's part number:		Manufacturer's serial number:
Measurement limit:		Date introduced:
Location:		
Instructions for use:		
Calibration frequency:		Signature of person responsible for calibration:
	CALIBRATION RECORD	
Calibration date	Calibration results	Calibrated by

Figure 4.1 *Typical format for instrument record sheets*

related measurements must be calibrated from time to time against a working standard instrument. As this working standard instrument is one which is kept by the instrumentation department of a company for calibration duties, and for no other purpose, then it can be assumed that it will maintain its accuracy over a reasonable period of time because use-related deterioration in accuracy is largely eliminated. However, over the longer term, the characteristics of even such a standard instrument will drift, mainly because of ageing effects in components within it. Over this longer term, therefore, a programme must be instituted for calibrating this working standard instrument against one of yet higher accuracy at appropriate intervals of time. The instrument used for calibrating working standard instruments is known as a secondary reference standard. This must obviously be a very well-engineered instrument which gives high accuracy and is stabilized against drift in its performance with time. This implies that it will be an expensive instrument to buy. It also requires that the environmental conditions in which it is used are carefully controlled in respect of ambient temperature, humidity, etc.

When the working standard instrument has been calibrated by an authorized standards laboratory, a calibration certificate will be issued (see Namas Document B 5103, 1985). This will contain at least the following information:

1 The identification of the equipment calibrated.
2 The calibration results obtained.
3 The measurement uncertainty.
4 Any use limitations on the equipment calibrated.
5 The date of calibration.
6 The authority under which the certificate is issued.

The establishment of a company standards laboratory to provide a calibration facility of the required quality is economically viable only in the case of very large companies where large numbers of instruments need to be calibrated across several factories. In the case of small to medium size companies, the cost of buying and maintaining such equipment is not justified. Instead, they would normally use the calibration service provided by various companies which specialize in offering a standards laboratory. What these specialist calibration companies effectively do is to share out the high cost of providing this highly accurate but infrequently used calibration service over a large number of companies. Such standards laboratories are closely monitored by national standards organizations (see ISO guide 25, 1982; BS 6460, 1983).

4.4 Validation of standards laboratories

In the United Kingdom, the appropriate national standards organization for validating standards laboratories is the National Physical Laboratory (NPL) and in the United States of America, the equivalent body is the National Bureau of Standards. The NPL has established a National Measurement Accreditation Service (NAMAS) which monitors both instrument calibration and mechanical testing laboratories. The formal structure for accrediting instrument calibration standards laboratories is known as the British Calibration Service (BCS), and that for accrediting testing facilities is known as the National Testing Laboratory Accreditation Scheme (NATLAS).

Although each country has its own structure for the maintenance of standards, each of these different frameworks tends to be equivalent in its effect. To achieve confidence in the goods and services which move across national boundaries, international agreements have established the equivalence of the different accreditation schemes in existence. As a result, NAMAS and the similar schemes operated by France, Germany, Italy, the United States, Australia and New Zealand enjoy mutual recognition.

The British Calibration Service lays down strict conditions which a standards laboratory has to meet before it is approved. These conditions control laboratory management, environment, equipment and documentation. The person appointed as head of the laboratory must be suitably qualified, and independence of operation of the laboratory must be guaranteed. The management structure must be such that any pressure to rush or skip calibration procedures for production reasons can be resisted. As far as the laboratory environment is concerned, proper temperature and humidity control must be provided, and high standards of cleanliness and housekeeping must be maintained. All equipment used for calibration purposes must be maintained to reference standards, and supported by calibration certificates which establish this traceability. Finally, full documentation must be maintained. This should describe all calibration procedures, maintain an index system for recalibration of equipment, and include a full inventory of apparatus and traceability schedules. Having met these conditions, a standards laboratory becomes an accredited laboratory for providing calibration services and issuing calibration certificates. This accreditation is reviewed at approximately twelve monthly intervals to ensure that the laboratory is continuing to satisfy the conditions laid down for approval.

4.5 Primary reference standards

Primary reference standards, as listed in Table 1.1, describe the highest level of accuracy that is achievable in the measurement of any particular physical quantity. All items of equipment used in standards laboratories as secondary reference standards have to be calibrated themselves against primary reference standards at appropriate intervals of time. This procedure is acknowledged by the issue of a calibration certificate in the usual way. National standards organizations maintain primary reference standards and suitable facilities for calibration against them within their own laboratories. In certain cases, such primary reference standards can be located outside national standards organizations. For instance, the primary reference standard for dimension measurement is defined by the wavelength of the orange–red line of krypton light, and it can therefore be realized in any laboratory equipped with an interferometer.

In certain cases, for example, the measurement of viscosity, such primary reference standards are not available and reference standards for calibration are achieved by collaboration between several national standards organizations who perform measurements on identical samples under controlled conditions (see BS 5497, 1987; ISO 5725, 1986).

4.6 Traceability

What has emerged from the foregoing discussion is that calibration has a chain-like structure in which every instrument in the chain is calibrated against a more accurate instrument immediately above it in the chain, as shown in Figure 4.2. All of the elements in the calibration chain must be known so that the calibration of process instruments at the bottom of the chain is traceable to the fundamental measurement standards.

This knowledge of the full chain of instruments involved in the calibration procedure is known as traceability, and is specified as a mandatory requirement in satisfying the BS.EN.ISO 9000 standard. Documentation must exist which shows that

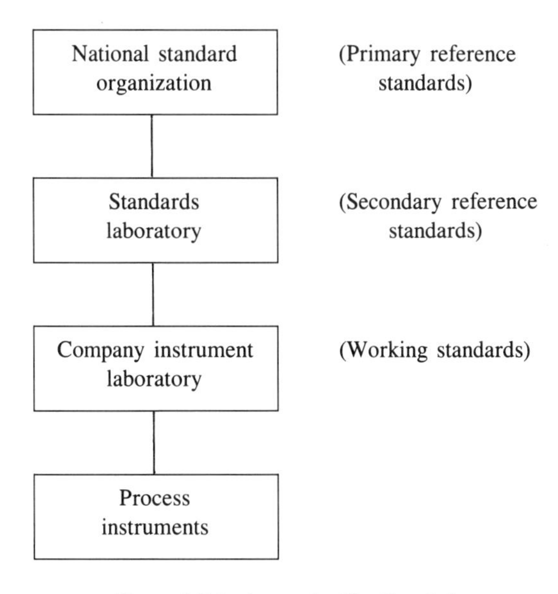

Figure 4.2 *Instrument calibration chain*

process instruments are calibrated by standard instruments which are linked by a chain of increasing accuracy back to national reference standards. There must be clear evidence to show that there is no break in this chain.

To illustrate a typical calibration chain, consider the calibration of micrometers as shown in Figure 4.3. A typical shop-floor micrometer has an uncertainty, i.e. inaccuracy, of less than 1 in 10^4. These would normally be calibrated in the instrumentation department or standards laboratory of a company against laboratory standard gauge blocks with a typical uncertainty of less than 1 in 10^5. A specialist calibration service company would provide facilities for calibrating these laboratory standard gauge blocks against reference grade gauge blocks with a typical uncertainty of less than 1 in 10^6. More accurate calibration equipment still is provided by national standards organizations. The National Physical Laboratory maintains two sets of standards for this type of calibration, a working standard and a primary standard. Spectral lamps are used to provide a working reference standard with an uncertainty of less than 1 in 10^7. The primary standard is provided by an iodine-stabilized helium–neon laser which has a specified uncertainty of less than 1 in 10^9. All of the links in this calibration chain must be shown in any documentation which describes the use of micrometers in making quality-related measurements.

4.7 Documentation in the workplace

An essential element in the maintenance of measurement systems and the operation of

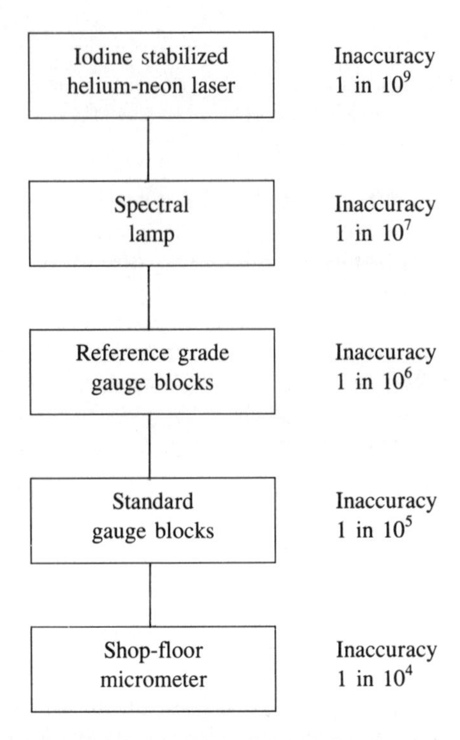

Figure 4.3 *Typical calibration chain for micrometers*

calibration procedures is the provision of full documentation. This must give a full description of the measurement requirements throughout the workplace, the instruments used, and the calibration system and procedures operated. Individual calibration records for each instrument must be included within this. This documentation is a necessary part of the Quality Manual, although it may physically exist as a separate volume if this is more convenient. An overriding constraint on the style in which the documentation is presented is that it should be simple and easy to read. This is often greatly facilitated by a copious use of appendices.

The starting point in the documentation must be a statement of what measurement limits have been defined for each measurement system documented. Such limits are established by balancing the costs of improved accuracy against customer requirements, and also with regard to what overall quality level has been specified in the Quality Manual. The technical procedures required for this, which involve assessing the type and magnitude of relevant measurement errors, are described in Chapter 3. It is customary to express the final measurement limit calculated as ±2 standard deviations, i.e. within 95% confidence limits (see Chapter 3 for explanation of these terms).

The instruments specified for each measurement situation must be listed next. This list must be accompanied by full instructions about the proper use of the instruments concerned. These instructions will include details about any environmental control or other special precautions which must be taken to ensure that the instruments provide measurements of sufficient accuracy to meet the measurement limits defined. The proper training courses appropriate to plant personnel who will use the instruments must be specified.

Having disposed of the question about what instruments are used, the documentation must go on to cover the subject of calibration. Full calibration is not applied to every measuring instrument used in a workplace because BS.EN.ISO 9000 acknowledges that formal calibration procedures are not necessary for some equipment where it is uneconomic or technically unnecessary because the accuracy of the measurement involved has an insignificant effect on the overall quality target for a product. However, any equipment which is excluded from calibration procedures in this manner must be specified as such in the documentation. Identification of equipment which is in this category is a matter of informed judgement.

For instruments which are the subject of formal calibration, the documentation must specify what standard instruments are to be used for the purpose and define a formal procedure of calibration. This procedure must include instructions for the storage and handling of standard calibration instruments and specify the required environmental conditions under which calibration is to be performed. Where a calibration procedure for a particular instrument uses published standard practices, it is sufficient to include reference to that standard procedure in the documentation rather than to reproduce the whole procedure. Whatever calibration system is established, a formal review procedure must be defined in the documentation which ensures its continued effectiveness at regular intervals. The results of each review must also be documented in a formal way.

A standard format for the recording of calibration results should be defined in the documentation. A separate record must be kept for every instrument present in the workplace which includes details of the instrument's description, the required calibration frequency, the date of each calibration and the calibration results on each occasion. Where appropriate, the documentation must also define the manner in which calibration results are to be recorded on the instruments themselves.

The documentation must specify procedures which are to be followed if an instrument is found to be outside the calibration limits. This may involve adjustment, redrawing its scale or withdrawing an instrument, depending upon the nature of the discrepancy and the type of instrument involved. Instruments withdrawn will either be repaired or scrapped. In the case of withdrawn instruments, a formal procedure for marking them as such must be defined to prevent them being accidentally put back into use.

Two other items must also be covered by the calibration document. The traceability of the calibration system back to national reference standards must be defined and supported by calibration certificates (see section 4.3). Training procedures must also be documented, specifying the particular training courses to be attended by various personnel and what, if any, refresher courses are required.

All aspects of these documented calibration procedures will be given consideration as part of the periodic audit of the quality control system which calibration procedures are instigated to support. Whilst the basic responsibility for choosing a suitable interval between calibration checks rests with the engineers responsible for the instruments concerned, the quality system auditor will need to see the results of tests which show that the calibration interval has been chosen correctly and that instruments are not going outside allowable measurement uncertainty limits between calibrations. Particularly important in such audits will be the existence of procedures which are instigated in response to instruments found to be out of calibration. Evidence that such procedures are effective in avoiding degradation in the quality assurance function will also be required.

References and further reading

British Standards Society, *The operation of a company standards department*, 1979. British Standards Society, London.

BS 5497, *Guide for the Determination of Repeatability and Reproducibility for a Standard Test Method by Inter-laboratory Tests*,1987. British Standards Institution, London.

BS 6460, *Accreditation of Testing Laboratories*,1983. British Standards Institution, London.

BS.EN.ISO 9000, *Quality Management and Quality Assurance Standards,* 1994. British Standards Institution, London.

ISO Guide 25, *General Requirements for the Technical Competence of Testing Laboratories*, 1982. International Organization for Standards, Geneva.

ISO 5725, *Precision of Test Methods–Determination of Repeatability and Reproducibility by Inter-laboratory Tests*, 1986. International Organization for Standards, Geneva.

NAMAS Document B 5103, *Certificates of Calibration*, 1985. NAMAS Executive, National Physical Laboratory, Middlesex.

CHAPTER 5

Signal processing, manipulation and transmission

Signal processing is concerned with improving the quality of the reading or signal at the output of a measurement system. The form which signal processing takes depends on the nature of the raw output signal from a measurement transducer. Procedures of signal amplification, signal attenuation, signal linearization, bias removal and signal filtering are applied according to the form of correction required in the raw signal. Noise-free transmission of the signal between remote transducers and the usage point of signals is also an important target in measurement systems, and appropriate means for this are discussed in the final section of this chapter.

The implementation of signal processing procedures can be carried out either by analog techniques or digitally on a computer. Analog signal processing involves the use of various electronic circuits, usually built around the operational amplifier, whereas digital signal processing uses software modules on a digital computer to condition the input measurement data. Digital signal processing is inherently more accurate than analog techniques, but this advantage is greatly reduced in the case of measurements coming from analog transducers, which have to be converted by an analog to digital converter prior to digital processing, thereby introducing conversion errors. Digital processing is also much slower than analog processing.

For the purposes of explaining the procedures involved, this chapter mainly describes analog signal processing and concludes with a relatively brief discussion of the equivalent digital signal processing techniques. One particular reason for this method of treatment is that some prior analog signal conditioning is often necessary even when the major part of the signal processing is carried out digitally.

5.1 Signal amplification

Signal amplification is carried out when the typical signal output level of a measurement transducer is considered to be too low. Amplification by analog means is carried out by an operational amplifier. This is normally required to have a high input impedance so that its loading effect on the transducer output signal is minimized. In some circumstances, such as when amplifying the output signal from accelerometers and some optical detectors, the amplifier must also have a high frequency response, to avoid distortion of the output reading.

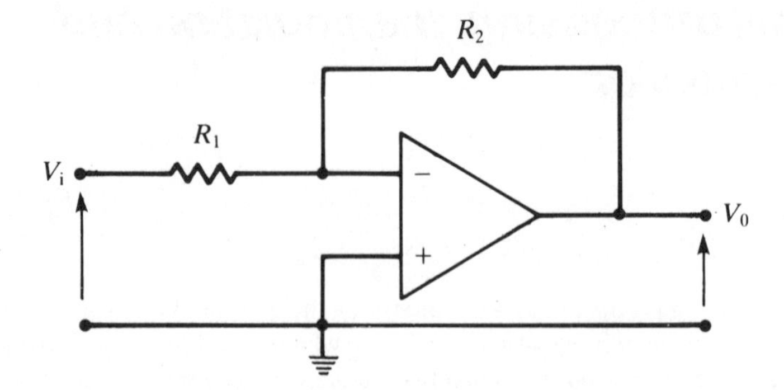

Figure 5.1 *Operational amplifier connected for signal amplification*

The operational amplifier is an electronic device which has two input terminals and one output terminal, the two inputs being known as the inverting input and non-inverting input respectively. When connected as shown in Figure 5.1, it provides signal amplification. The raw (unprocessed) signal V_i is connected to the inverting input through a resistor R_1 and the non-inverting input is connected to ground. A feedback path is provided from the output terminal through a resistor R_2 to the inverting input terminal. Assuming ideal operational amplifier characteristics, the processed signal V_o at the output terminal is then related to the voltage V_i at the input terminal by the expression:

$$V_o = \frac{-R_2 V_i}{R_1} \tag{5.1}$$

The amount of signal amplification is therefore defined by the relative values of R_1 and R_2. This ratio between R_1 and R_2 in the amplifier configuration is often known as the amplifier gain or closed-loop gain. If, for instance, $R_1 = 1\,M\Omega$ and $R_2 = 10\,M\Omega$, an amplification factor of 10 is obtained (i.e. gain = 10). It is important to note that, in this standard way of connecting the operational amplifier (often known as the inverting configuration), the polarity (sign) of the processed signal is inverted. This can be corrected if necessary by feeding the signal through a further amplifier configured for unity gain ($R_1 = R_2$). This inverts the signal again and returns it to its original polarity.

5.1.1 Instrumentation amplifier

For some applications requiring the amplification of very low-level signals, a special type of amplifier known as an *instrumentation amplifier* is used. This consists of a circuit containing three standard operational amplifiers, as shown in Figure 5.2. The advantage of the instrumentation amplifier compared with a standard operational amplifier is that its differential input impedance is much higher. In consequence, its

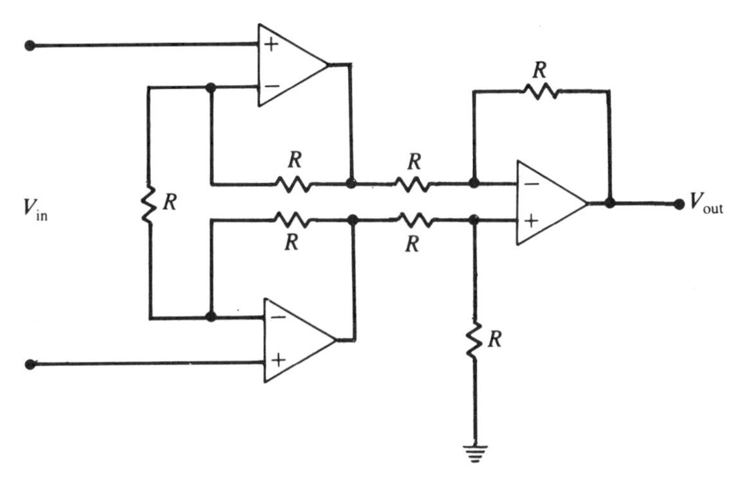

Figure 5.2 *Instrumentation amplifier*

common mode rejection capability[1] is much better. This means that, if a twisted-wire pair is used to connect a transducer to the differential inputs of the amplifier, any induced noise will contaminate each wire equally and will be rejected by the common mode rejection capacity of the amplifier.

5.2 Signal attenuation

One method of attenuating signals by analog means is to use a potentiometer connected in a voltage-dividing circuit, as shown in Figure 5.3. For the potentiometer wiper positioned a distance of X_w along the resistance element of total length X_t, the voltage level of the processed signal V_o is related to the voltage level of the input signal V_i by the expression:

$$V_o = \frac{X_w V_i}{X_t}$$

A major problem with potentiometers in many circuits is that the output can be affected by the impedance of the device (or circuit) connected to its output terminals. An alternative device to the potentiometer for signal attenuation is the operational amplifier. This is connected in the same way as the amplifier shown in Figure 5.1, but R_1 is chosen to be greater than R_2. Equation (5.1) is still valid and therefore, if R_1 is chosen to be 10 MΩ and R_2 as 1 MΩ, an attenuation factor of 10 is achieved (gain = 0.1). Use of an operational amplifier as an attenuating device is a more expensive

[1] Common mode rejection ratio describes the ability of the amplifier to reject equal-magnitude signals that appear on both of its inputs.

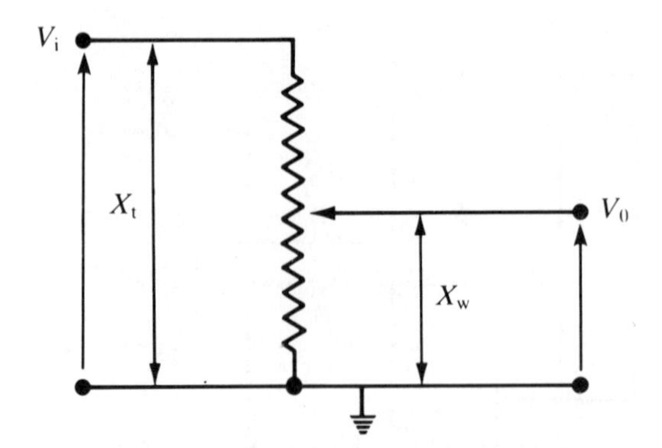

Figure 5.3 *Potentiometer in voltage-dividing circuit*

solution than using a potentiometer, particularly because it is an active device and needs a power supply. However, unlike the potentiometer, it is relatively unaffected by the device or circuit connected to its output terminals.

5.3 Signal linearization

Several types of transducer used in measuring instruments have an output which is a non-linear function of the measured input quantity. In many cases, this non-linear signal can be converted to a linear one by special operational amplifier configurations which have an equal and opposite non-linear relationship between the amplifier input and output terminals.

For example, light intensity transducers typically have an exponential relationship between the output signal and the input light intensity of the form:

$$V_o = K e^{-\alpha Q} \tag{5.2}$$

where Q is the light intensity, V_o is the voltage level of the output signal, and K and α are constants.

If a diode is placed in the feedback path between the input and output terminals of the amplifier as shown in Figure 5.4, the relationship between the amplifier output voltage V_2 and input voltage V_1 is given by:

$$V_o = C \log_e(V_i) \tag{5.3}$$

where C is a constant.

If the output of the light intensity transducer with a characteristic given by equation (5.2) is conditioned by an amplifier with a characteristic given by equation (5.3), the voltage level of the processed signal is given by:

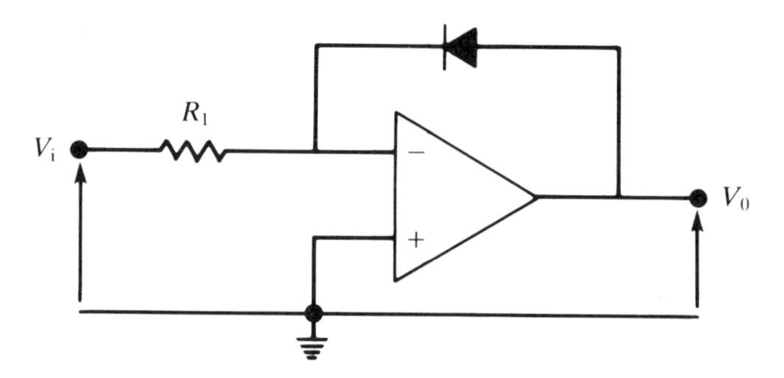

Figure 5.4 *Operational amplifier connected for signal linearization*

$$V_o = C \log_e(K) - \alpha CQ \tag{5.4}$$

Expression (5.4) shows that the output signal now varies linearly with light intensity Q but with an offset of $C \log_e(K)$. This offset would normally be removed by further signal conditioning, as described below.

5.4 **Bias removal**

Sometimes, either because of the nature of the measurement transducer itself, or as a result of other signal-conditioning operations as previously described in section 5.3, a bias exists in the output signal. This can be expressed mathematically for a physical quantity x and measurement signal y as:

$$y = Kx + C \tag{5.5}$$

where C represents a bias in the output signal which needs to be removed by signal processing. Analog processing consists of using an operational amplifier connected in a differential amplification mode, as shown in Figure 5.5. Referring to this circuit, for $R_1 = R_2$ and $R_3 = R_4$, the output V_o is given by:

$$V_o = (R_3/R_1)(V_c - V_i) \tag{5.6}$$

where V_i is the unprocessed measurement signal equal to $(Kx + C)$ and V_c is the output voltage from a potentiometer supplied by a known reference voltage V_{ref}, which is set such that $V_c = C$. Now, substituting these values for V_i and V_c into equation (5.6), y can be written as:

$$y = K'x \tag{5.7}$$

where the new constant K' is related to K according to $K' = -K(R_3/R_1)$. It is clear that the bias has been successfully removed and equation (5.7) is now a linear relationship between the measurement signal y and the measured quantity x.

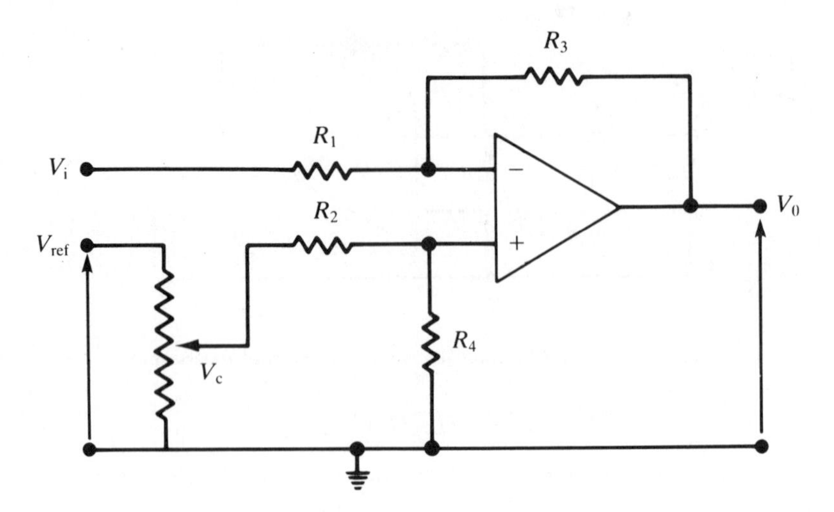

Figure 5.5 *Operational amplifier connected in differential amplification mode*

5.5 Signal filtering

Signal filtering consists of processing a signal to remove a certain band of frequencies within it. The band of frequencies removed can be at the low-frequency end of the frequency spectrum, at the high frequency end, at both ends, or in the middle of the spectrum. Filters to perform each of these operations are known respectively as low-pass filters, high-pass filters, band-pass filters and band-stop filters. All such filtering operations can be carried out by either analog or digital methods.

The result of filtering can be readily understood if the analogy with a procedure such as sieving soil particles is considered. Suppose that a sample of soil A is passed through a system of two sieves of differing meshes such that the soil is divided into three parts, B, C and D, consisting of large, medium and small particles, as shown in Figure 5.6. Suppose that the system also has a mechanism for delivering one or more of the separated parts, B, C and D, as the system output. If the graded soil output consists of parts C and D, the system is behaving as a low-pass filter and rejecting large particles, whereas if it consists of parts B and C, the system is behaving as a high-pass filter and rejecting small particles. Other options are to deliver just part C (band-pass filter mode) or parts B and D together (band-stop filter mode). As any gardener knows, however, such perfect sieving is not achieved in practice and any form of graded soil output always contains a few particles of the wrong size.

Signal filtering consists of selectively passing or rejecting low-, medium- and high-frequency signals from the frequency spectrum of a general signal. The range of frequencies passed by a filter is known as the *pass band*, the range not passed is known as the *stop band*, and the boundary between the two ranges is known as the *cut-off frequency*. To illustrate this, consider a signal whose frequency spectrum is such that

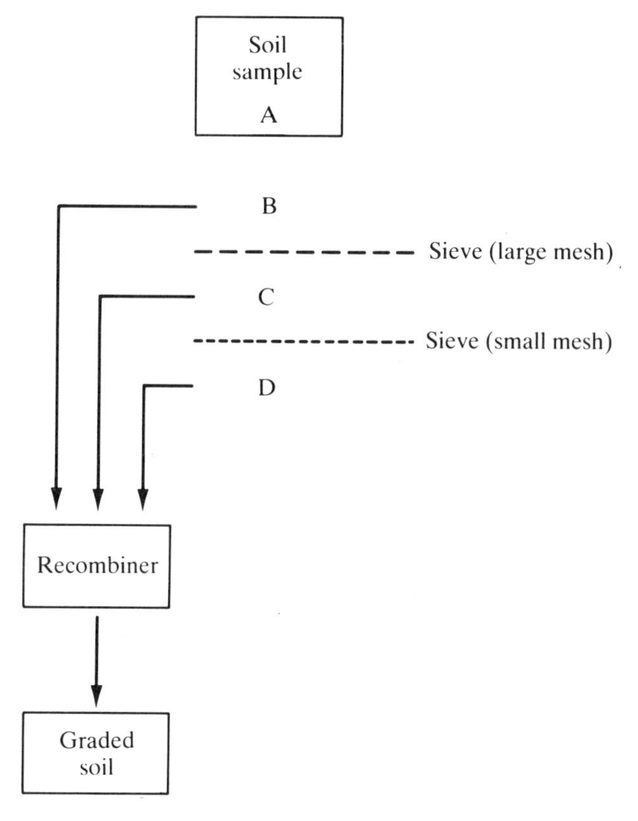

Figure 5.6 *Soil sieving analogy of signal filtering*

all frequency components in the frequency range from zero to infinity have equal magnitude. If this signal is applied to an ideal filter, then the outputs for a low-pass filter, high-pass filter, band-pass filter and band-stop filter respectively are as shown in Figure 5.7. Note that for the latter two types, the bands are defined by a pair of frequencies rather than by a single cut-off frequency.

Just as in the case of the soil sieving analogy presented above, the signal-filtering mechanism is not perfect, with unwanted frequency components not being erased completely but only attenuated by varying degrees. Thus, the filtered signal always retains some components in the unwanted frequency range, but these are of relatively low magnitude. There is also a small amount of attenuation of frequencies within the pass band, which increases as the cut-off frequency is approached. Figure 5.8 shows the typical output characteristics of a practical constant-k filter[2] designed

[2] 'Constant-k' is a term used to describe a common class of passive filters, as discussed in the following section.

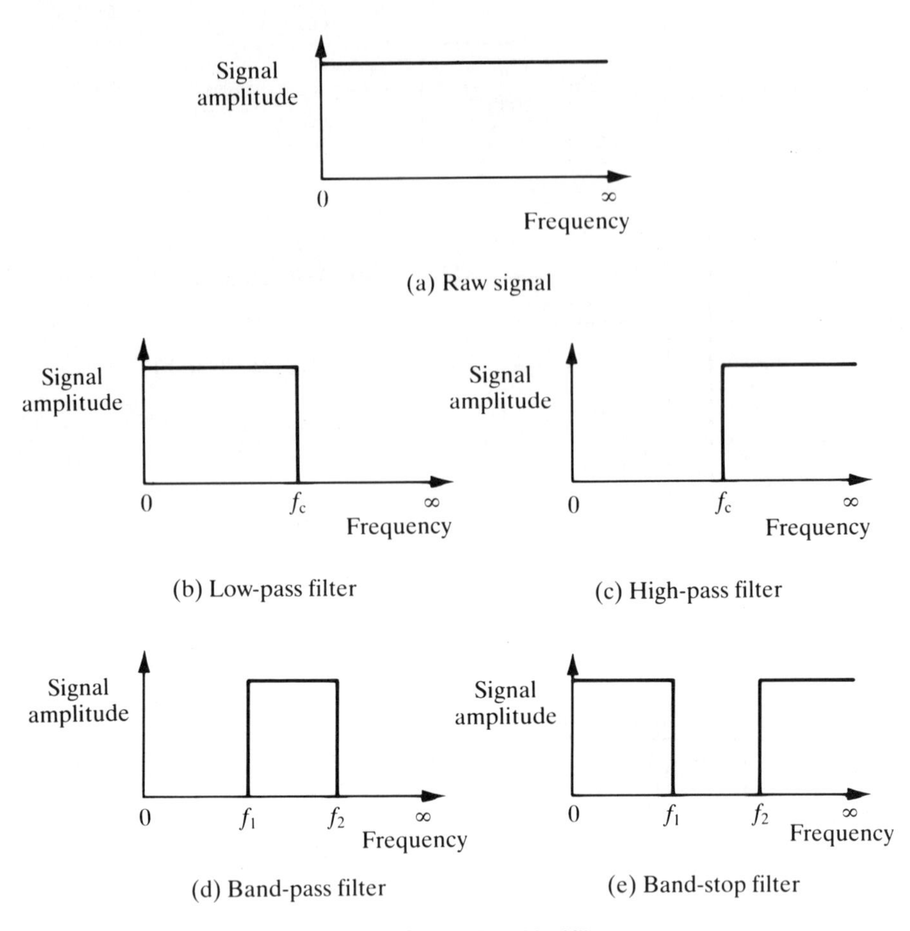

Figure 5.7 *Outputs from ideal filters*

respectively for low-pass, high-pass, band-pass and band-stop filtering. Filter design is concerned with trying to obtain frequency rejection characteristics which are as close to the ideal as possible. However, improvement in characteristics is only achieved at the expense of greater complexity in the design. The filter chosen for any given situation is therefore a compromise between performance, complexity and cost.

In the majority of measurement situations, the physical quantity being measured has a value which is either constant or only changing slowly with time. In these circumstances, the most common types of signal corruption are high-frequency noise components, and the type of signal processing element required is a low-pass filter. In a few cases, the measured signal itself has a high frequency, for instance when mechanical vibrations are being monitored, and the signal processing required is the application of a high-pass filter to attenuate low-frequency noise components. Band-stop filters can be used where a measurement signal is corrupted by noise at a particular

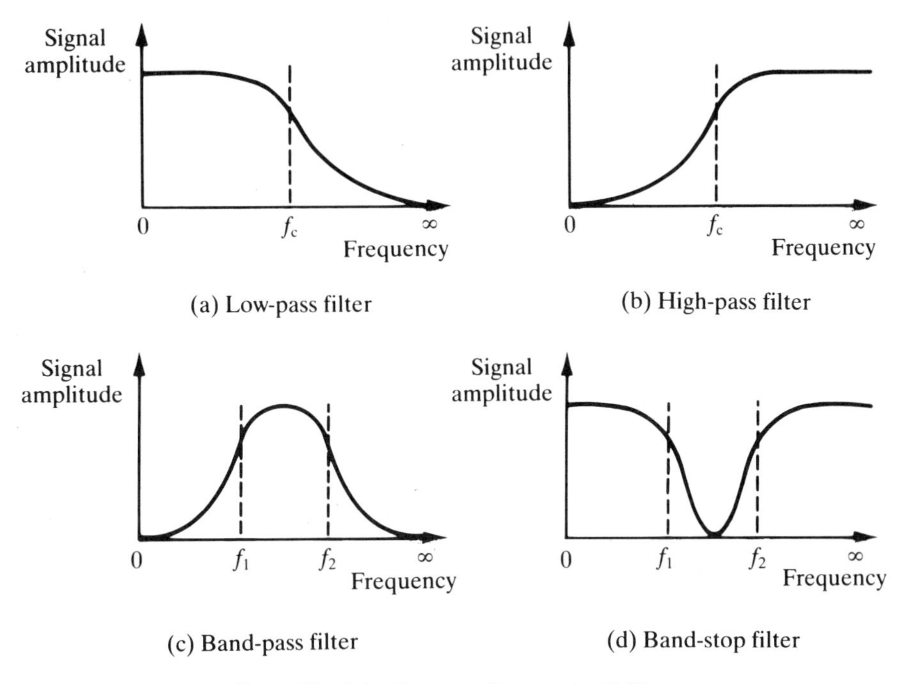

Figure 5.8 *Output from practical constant-k filters*

frequency. Such noise is frequently due to mechanical vibrations or proximity of the measurement circuit to other electrical apparatus.

Both passive and active analog filter implementations of the four types of filter identified are considered below. The equivalent digital filters are discussed later in section 5.7.

5.5.1 **Passive analog filters**

The detailed design of passive filters is a subject of some considerable complexity which is outside the scope of this book. In the following treatment, the major formulae appropriate to the design of filters are quoted without derivation. The derivation can be found elsewhere (Blinchikoff, 1976; Skilling, 1967; Williams, 1963).

Simple passive filters consist of a network of impedances, such as those labelled as Z_1 and Z_2 in Figure 5.9(a). So that there is no dissipation of energy in the filter, these impedances should ideally be pure reactances (capacitors or resistance-less inductors). In practice, however, it is impossible to manufacture inductors that do not have a small resistive component and so this ideal cannot be achieved. Indeed, readers familiar with radio receiver design will be aware of the existence of filters consisting only of resistors and capacitors. These only have a mild filtering effect which is useful in radio tone controls but not relevant to the signal-processing

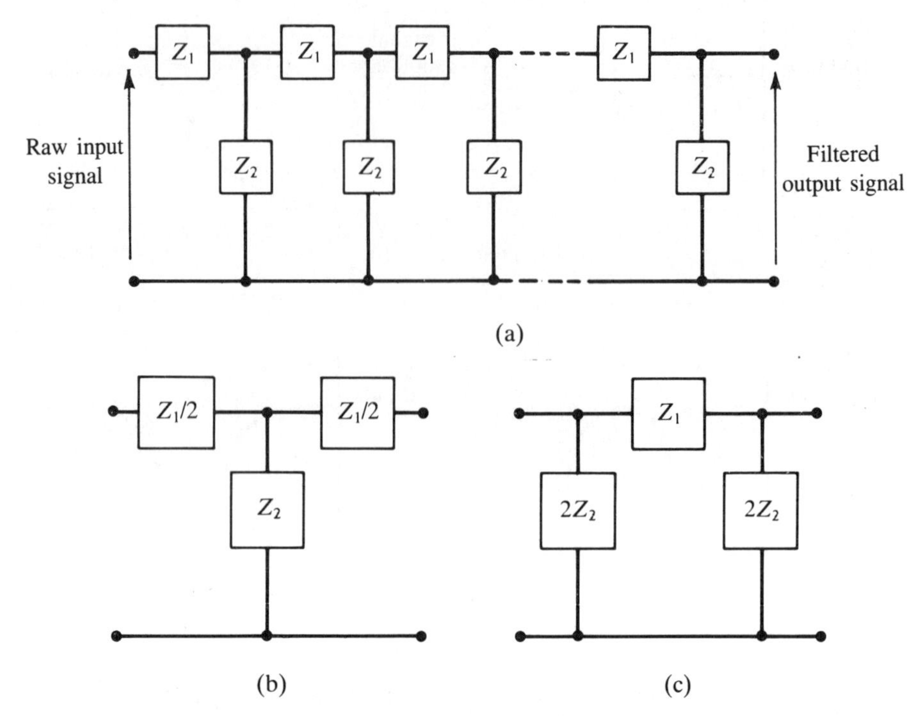

Figure 5.9 *(a) Simple passive filter; (b) T-section; (c) π-section*

requirements discussed in this chapter. Such filters are therefore not considered further.

Each element of the network shown in Figure 5.9(a) can be represented by either a T-section or a π-section as shown in (b) and (c), respectively. To obtain proper matching between filter sections, it is necessary for the input impedance of each section to be equal to the output load impedance for that section. This value of impedance is known as the characteristic impedance (Z_o). For a T-section of filter, the characteristic impedance is calculated from:

$$Z_o = \sqrt{Z_1 \cdot Z_2[1 + (Z_1/4Z_2)]} \tag{5.8}$$

The frequency attenuation characteristics of the filter can be determined by inspecting this expression for Z_o. Frequency values for which Z_o is real lie in the pass band of the filter and frequencies for which Z_o is imaginary lie in its stop band.

Let $Z_1 = j\omega L$ and $Z_2 = 1/j\omega C$, where L is an inductance value, C is a capacitance value and ω is the angular frequency in rad/s which is related to the frequency f in Hz according to $\omega = 2\pi f$. Substituting these values into the equation (5.8), we obtain:

$$Z_o = \sqrt{(L/C) \cdot (1 - 0.25\omega^2 LC)}$$

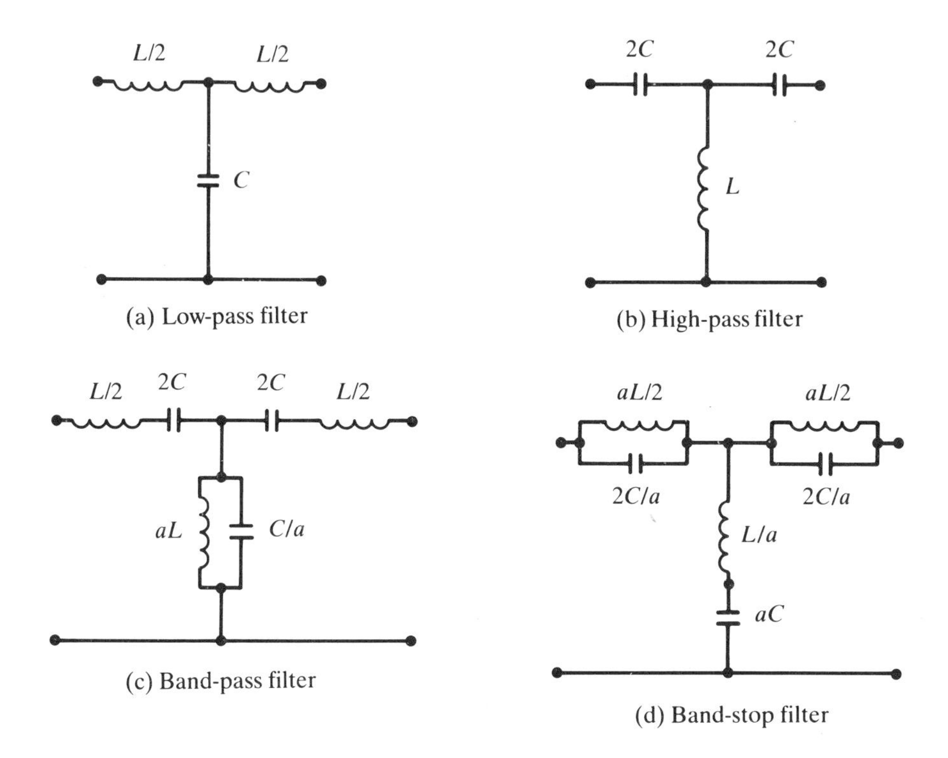

Figure 5.10 *Circuit components for passive filter T-sections*

For frequencies where $\omega < \sqrt{4/LC}$, Z_o is real, and for higher frequencies, Z_o is imaginary. These values of impedance therefore give a *low-pass filter*, as shown in Figure 5.10(a), with cut-off frequency f_c given by:

$$f_c = (\omega_c/2\pi) = \frac{1}{2\pi\sqrt{LC}}$$

A *high-pass* filter (see Figure 5.10(b)) can be synthesized with exactly the same cut-off frequency if the impedance values chosen are: $Z_1 = 1/j\omega C$ and $Z_2 = j\omega L$.

A point worthy of note in both these examples is that the product $Z_1 \cdot Z_2$ could be represented by a constant *k* which is independent of frequency. Because of this, such filters are known by the name of constant-*k* filters.

A constant-*k* *band-pass* filter (see Figure 5.10(c)) can be realized with the following choice of impedance values, where *a* is a constant and other parameters are as before:

$$Z_1 = j\omega L + \frac{1}{j\omega C} \qquad Z_2 = \frac{(j\omega aL)(a/j\omega C)}{j\omega aL + (a/j\omega C)}$$

The frequencies f_1 and f_2 defining the end of the pass band are most easily expressed in terms of a frequency f_0 in the centre of the pass band. The corresponding equations are:

$$f_0 = \frac{1}{2\pi(LC)^{1/2}}$$
$$f_1 = f_0[(1+a)^{1/2} - a^{1/2}]$$
$$f_2 = f_0[(1+a)^{1/2} + a^{1/2}]$$

For a constant-*k* *band-stop* filter (see Figure 5.10(d)), the appropriate impedance values are:

$$Z_1 = \frac{(j\omega La)(a/j\omega C)}{j\omega La + (a/j\omega C)}$$
$$Z_2 = \frac{1}{a}\left(j\omega L + \frac{1}{j\omega C}\right)$$

The frequencies defining the ends of the stop band are again normally defined in terms of the frequency f_0 in the centre of the stop band:

$$f_0 = \frac{1}{2\pi(LC)^{1/2}}$$
$$f_1 = f_0\left(1 - \frac{a}{4}\right)$$
$$f_2 = f_0\left(1 + \frac{a}{4}\right)$$

As has already been mentioned, a practical filter does not eliminate frequencies in the stop band but merely attenuates them by a certain amount. The attenuation coefficient, α, at a frequency in the stop band, f, for a single T-section of a low-pass filter is given by:

$$\alpha = 2\cosh^{-1}[f/fc] \qquad (5.9)$$

The relatively poor attenuation characteristics are obvious if we evaluate this expression for a value of frequency close to the cut-off frequency given by $f = 2f_c$. Then $\alpha = 2\cosh^{-1}(2) = 2.64$. Further away from the cut-off frequency, for $f = 20f_c$, $\alpha = 2\cosh^{-1}(20) = 7.38$.

Improved attenuation characteristics can be obtained by putting several T-sections in cascade. If perfect matching is assumed then two T-sections give twice the attenuation of one section, i.e. at frequencies of $2f_c$ and $20f_c$, α for two sections would have a value of 5.28 and 14.76 respectively.

The discussion so far has assumed resistance-less inductances and perfect matching between sections. Such conditions cannot be achieved in practice and this has several consequences.

Inspection of the expression for the characteristic impedance (5.8) reveals frequency-dependent terms. Thus the condition that the load impedance is equal to the input impedance for a section is only satisfied at one particular frequency. The load impedance is normally chosen so that this frequency is 'well within the pass band'. This normally means, to satisfy this condition, choosing zero frequency for a low-pass filter and infinite frequency for a high-pass filter. Frequency-dependency is one of the reasons for the degree of attenuation in the pass band shown in the practical filter characteristics of Figure 5.8, the other reason being the presence of resistance in the inductors of the filter. The effect of this in a practical filter is that the value of α at the cut-off frequency is 1.414 whereas the value predicted theoretically for an ideal filter (equation 5.9) is zero. Cascading filter sections together increases this attenuation in the pass band as well as increasing attenuation of frequencies in the stop band.

This problem of matching successive sections in a cascaded filter seriously degrades the performance of constant-k filters and this has resulted in the development of other types such as m-derived and n-derived composite filters. These produce less attenuation within the pass band and greater attenuation outside it than constant-k filters, although this is only achieved at the expense of greater filter complexity and cost. The reader interested in further consideration of these is directed to consult one of the specialist texts recommended in the further reading section at the end of the chapter.

5.5.2 *Active analog filters*

In the foregoing discussion on passive filters, the two main problems noted were those of obtaining non-resistive inductors and of achieving proper matching between the signal source and load through the filter sections. A further problem is that the inductors required by passive filters are bulky and relatively expensive. Active filters overcome all of these problems and so are very popular for signal-processing duties.

The major component in an active filter is an electronic amplifier, with the filter characteristics being defined by a network of amplifier input and feedback components consisting of resistors and capacitors but no inductors. Active circuits to produce the four different types of filtering defined are illustrated in Figure 5.11. These particular circuits are all known as second order filters because the relationship between filter input and output is described by a second order differential equation.

As for passive filters, the design of active filters is a subject of considerable complexity. Simplified design formulae for the circuit parameters in Figure 5.11 which achieve specified cut-off frequencies and gains for the four standard filter types are given below. These formulae imply a fairly free choice of the R and C component values, but this is only because they omit consideration of the shape of the frequency attenuation curve, which in practice is important. Such consideration imposes additional equations which constrain the choice of the R and C parameters. For more information, the reader should consult advanced texts (e.g. Stephenson, 1985, or Hilburn and Johnson, 1973).

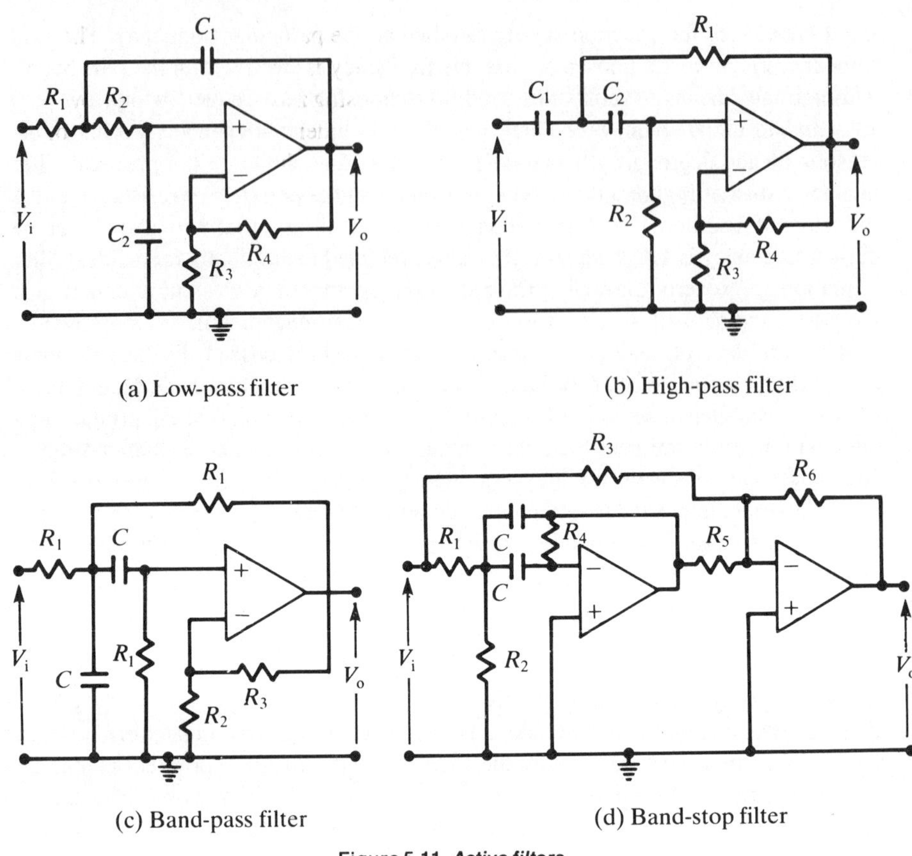

Figure 5.11 *Active filters*

1 *Parameters for low-pass filter*:

$$w_o = \text{cut-off frequency} = \sqrt{\frac{1}{R_1 R_2 C_1 C_2}}$$

G = filter gain (at d.c.) = $1 + R_4/R_3$

2 *Parameters for high-pass filter*:

$$w_o = \text{cut-off frequency} = \sqrt{\frac{1}{R_1 R_2 C_1 C_2}}$$

G = filter gain (at infinite frequency) = $1 + R_4/R_3$

3 *Parameters for band-pass filter*:

$$w_o = \text{centre frequency of pass band} = \sqrt{\frac{2}{R_1^2 C^2}}$$

G = filter gain (at frequency w_o) = $1 + R_3/R_2$

$$w_1 = w_0 - \frac{4-G}{2R_1C}; \quad w_2 = w_0 + \frac{4-G}{2R_1C}$$

where w_1 and w_2 are the frequencies at the ends of the pass band.

4 *Parameters for band-stop filter*:

$$w_o = \text{centre frequency of stop band} = \sqrt{\frac{1}{R_4C^2}\left(\frac{1}{R_1}+\frac{1}{R_2}\right)}$$

G = filter gain (at d.c. and high frequency) = $-R_6/R_3$

$$w_1 = w_0 - \frac{1}{R_4C}; \quad w_2 = w_0 + \frac{1}{R_4C}$$

where w_1 and w_2 are the frequencies at the ends of the stop band.

An alternative and simpler filter design procedure, which establishes parameters based on the above formulae, is to use graphs in which gains and cut-off frequencies are plotted for a range of parameter values, with suitable parameter values being chosen by inspection (see Hilburn and Johnson, 1973).

Filters with parameters derived from the above formulae provide active filters which are general purpose and suitable for most applications. However, many other design formulae exist for the parameters of filters with circuit structures as given in Figure 5.11, and these yield filters with special names and characteristics. Butterworth filters, for instance, optimize the pass band attenuation characteristics at the expense of stop band performance. Another form, Chebyshev filters, have very good stop band attenuation characteristics but poorer pass band performance. Again, the reader is referred to the specialist texts mentioned earlier for more information.

5.6 **Signal manipulation**

To complete the discussion on analog signal processing techniques, mention must also be made of certain other special-purpose devices and circuits used to manipulate signals. These are listed below.

5.6.1 ***Voltage-to-current conversion***

Many process-control systems use current to transmit signals rather than voltage. Hence, voltage-to-current conversion is important.

An operational amplifier circuit, as shown in Figure 5.12, is a suitable voltage-to-current converter in which the output current I is related to the input voltage V_{in} by the equation:

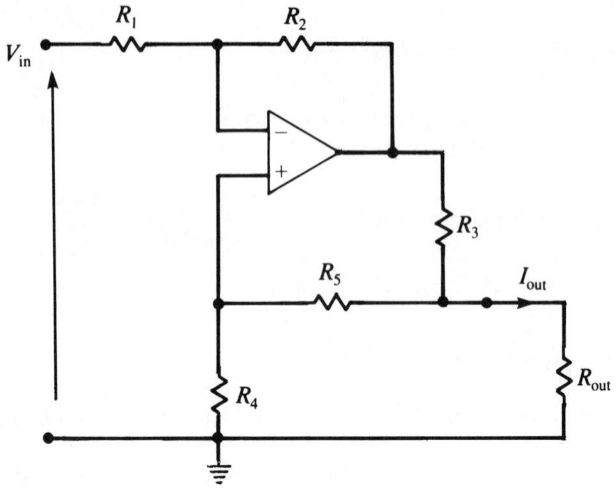

Figure 5.12 *Operational amplifier connected for voltage to current conversion*

$$I = -\frac{R_2}{R_1 R_3} V_{in}$$

If required, the amplifier gain can be reduced to lower the output current level by either increasing R_1 and R_3 or reducing R_2.

5.6.2 Current-to-voltage conversion

Current-to-voltage conversion is often required at the termination of transmission lines in process-control systems to change the transmitted currents back to voltages. Again, an operational amplifier, connected as shown in Figure 5.13, is suitable for this purpose. The output voltage V_{out} is simply related to the input current I by:

$$V_{out} = IR$$

5.6.3 Signal integration

Connected in the configuration shown in Figure 5.14, an operational amplifier is able to integrate the input signal V_{in} such that the output signal V_{out} is given by:

$$V_{out} = -\frac{1}{RC} \int V_{in} dt$$

This circuit is used whenever there is a requirement to integrate the output signal from a transducer.

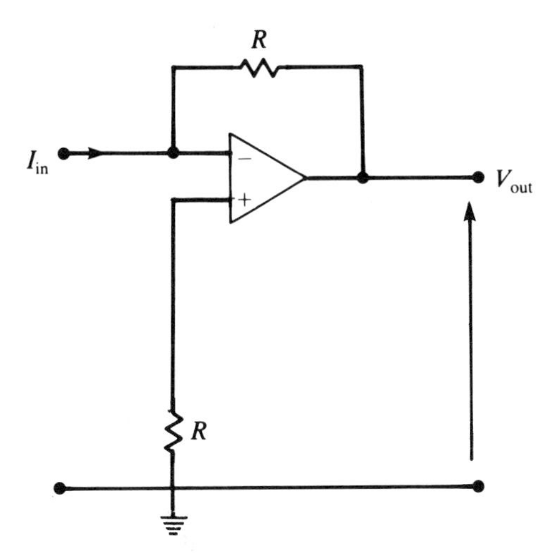

Figure 5.13 *Operational amplifier connected for current to voltage conversion*

5.6.4 **Voltage follower (pre-amplifier)**

The voltage follower, also known as a pre-amplifier, is a unity gain amplifier circuit with a short circuit in the feedback path, as shown in Figure 5.15, such that $V_{out} = V_{in}$.

It has a very high input impedance and its main application is to reduce the load on the measured system. It also has a very low output impedance which is very useful in some impedance-matching applications.

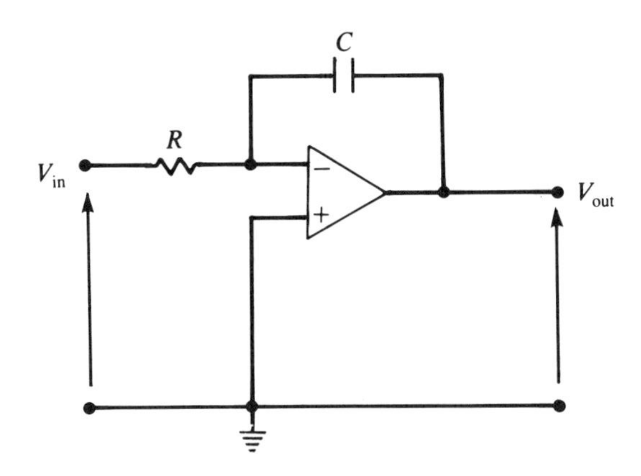

Figure 5.14 *Operational amplifier connected as integrating element*

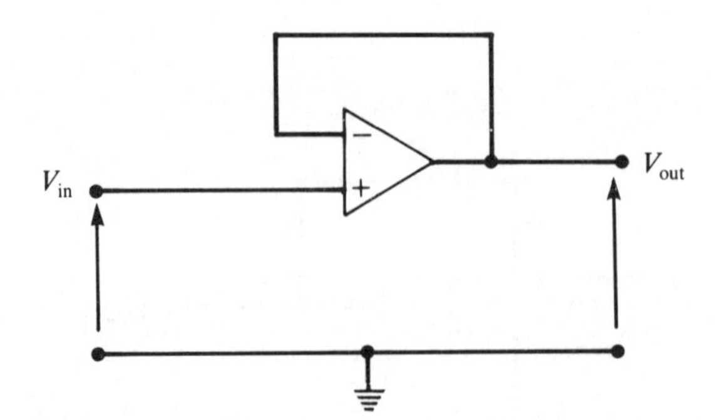

Figure 5.15 *Operational amplifier connected as voltage follower (pre-amplifier)*

5.6.5 **Voltage comparator**

The output of a voltage comparator switches between positive and negative values according to whether the difference between the two input signals to it is positive or negative. An operational amplifier connected as shown in Figure 5.16 gives an output which switches between positive and negative saturation levels according to whether $(V_1 - V_2)$ is greater or less than zero.

Alternatively, the voltage of a single input signal can be compared against positive and negative reference levels with the circuit shown in Figure 5.17.

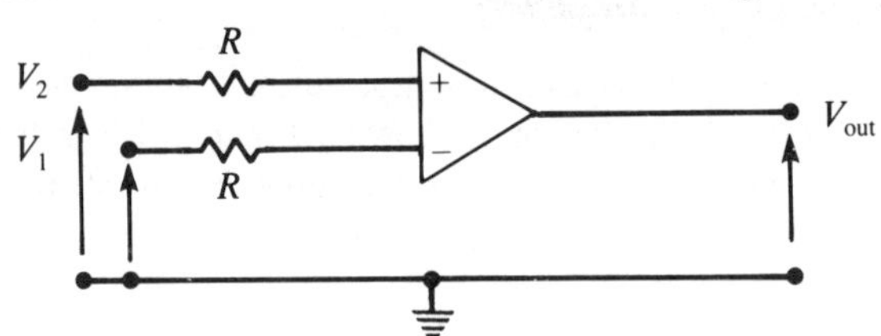

Figure 5.16 *Comparison between two voltage signals*

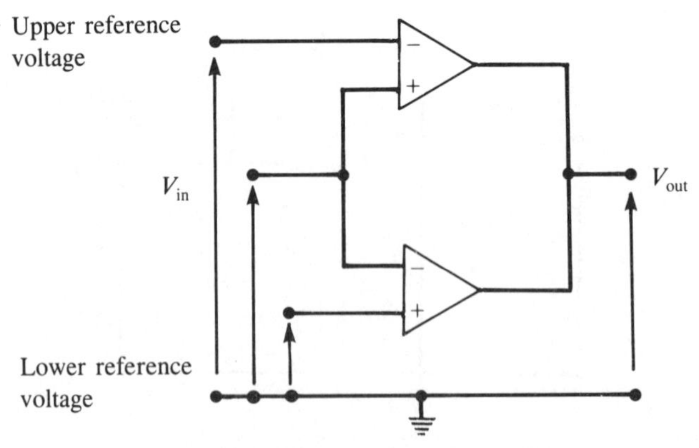

Figure 5.17 *Comparison of input signal against reference value*

In practice, operational amplifiers have drawbacks as voltage comparators for several reasons. These include propagation delays and non-compatibility between output voltage levels and industry-standard logic circuits, and slow recovery. In consequence, various other special-purpose integrated circuits have been developed for voltage comparison.

5.6.6 Phase-locked loop

The phase-locked loop, described in detail in section 7.8, is primarily a circuit for measuring the frequency of a signal. However, because the output waveform is a pure (i.e. perfectly clean) square wave at the same frequency as the input signal, irrespective of the amount of noise, modulation or distortion on the input signal, the phase-locked loop also finds application as a signal-processing element to clean up poor-quality signals.

5.6.7 Signal addition

The most common mechanism for summing two or more input signals is the use of an operational amplifier connected in signal-inversion mode, as shown in Figure 5.18. For input signal voltages V_1, V_2 and V_3, the output voltage V_o is given by:

$$V_o = -(V_1 + V_2 + V_3)$$

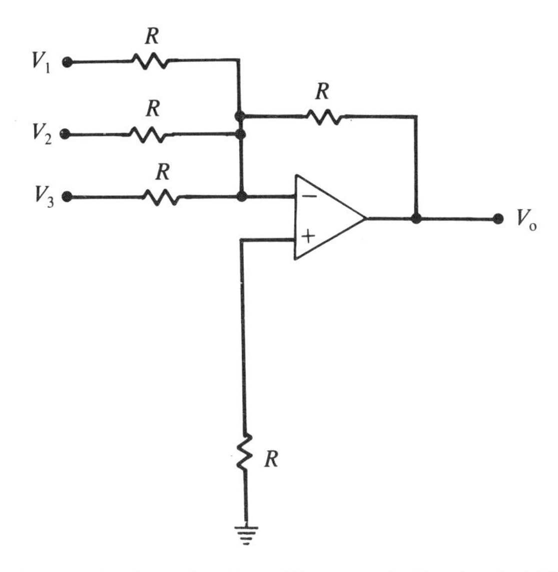

Figure 5.18 *Operational amplifier connected for signal addition*

5.6.8 *Signal multiplication*

Great care must be taken when choosing a signal multiplier because, while many circuits exist for multiplying two analog signals together, most of them are two-quadrant types which only work for signals of a single polarity, i.e. both positive or both negative. Such schemes are unsuitable for general analog signal processing, where the signals to be multiplied may be of changing polarity.

For analog signal processing, a four-quadrant multiplier is required. Two forms of such a multiplier are easily available, the Hall-effect multiplier and the translinear multiplier.

5.6.9 *Sample and hold circuit*

A sample and hold circuit is often an essential element at the interface between an analog instrument/transducer and an analog-to-digital converter. It holds the input signal at a constant level while the analog to digital conversion process is taking place and prevents the conversion errors which would probably result if variations in the measured signal were allowed to pass through to the converter. The operational amplifier circuit shown in Figure 5.19 provides this sample and hold function. The input signal is applied to the circuit for a very short time duration with switch S_1 closed and S_2 open, after which S_1 is opened and the signal level is then held until, when the next sample is required, the circuit is reset by closing S_2.

5.6.10 *Analog-to-digital conversion*

In most computer-controlled systems, there is a fundamental mismatch between the analog form of output data from instruments and transducers and the digital form of

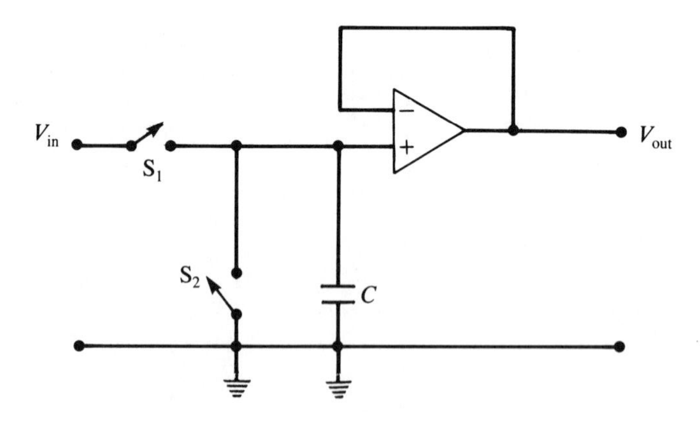

Figure 5.19 *Operational amplifier connected as 'sample and hold' circuit*

data required by a digital computer. This problem is solved by the provision of an analog-to-digital converter in the computer input interface.

5.6.11 *Digital-to-analog conversion*

Similarly in computer controlled systems, control actuators usually require the application of analog signals whereas the controlling computer has a digital output. This is overcome by providing a digital-to-analog converter in the computer output interface.

5.7 Digital signal processing

Digital techniques achieve much greater levels of accuracy in signal processing than equivalent analog methods. However, the time taken to process a signal digitally is much longer than that required to carry out the same operation by analog techniques, and the equipment required is more expensive. Some care is needed therefore in making the correct choice between digital and analog methods in a particular signal-processing application.

While digital signal-processing elements in a measurement system can exist as separate units, it is more usual to find them as an integral part of an intelligent instrument. However, their construction and mode of operation are the same irrespective of whether they exist physically as separate boxes or within an intelligent instrument.

The hardware aspect of a digital signal-processing element consists of a digital computer and analog interface boards. The actual form that signal processing takes depends on the software program executed by the processor. However, before consideration is given to this, some theoretical aspects of signal sampling need to be discussed.

As mentioned earlier, digital computers require signals to be in digital form whereas most instrumentation transducers have an output signal in analog form. Analog-to-digital conversion is therefore required at the interface between analog transducers and the digital computer. The procedure followed is to sample the analog signal at a particular moment in time and then convert the analog value to an equivalent digital one. This conversion takes a certain finite time, during which the analog signal can be changing in value. The next sample of the analog signal cannot be taken until the conversion of the last sample to digital form is completed. The representation within a digital computer of a continuous analog signal is therefore a sequence of samples whose pattern only approximately follows the shape of the original signal. This pattern of samples taken at successive, equal intervals of time is known as a discrete signal. The process of conversion between a continuous analog signal and a discrete digital one is illustrated for a sine wave in Figure 5.20.

The raw analog signal in Figure 5.20 has a frequency of approximately 0.75 cycles per second. With the rate of sampling shown, which is approximately eleven samples

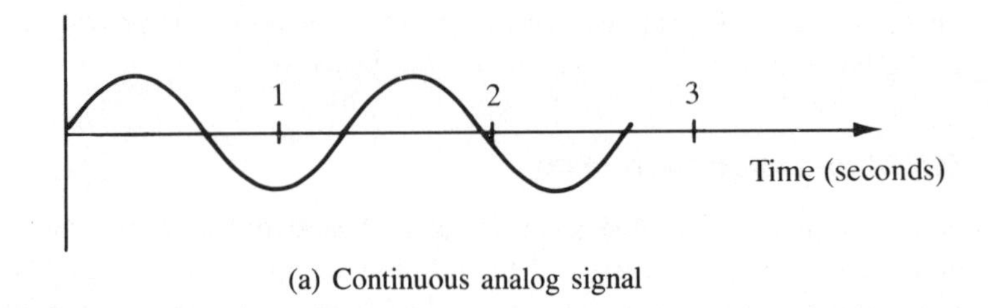

(a) Continuous analog signal

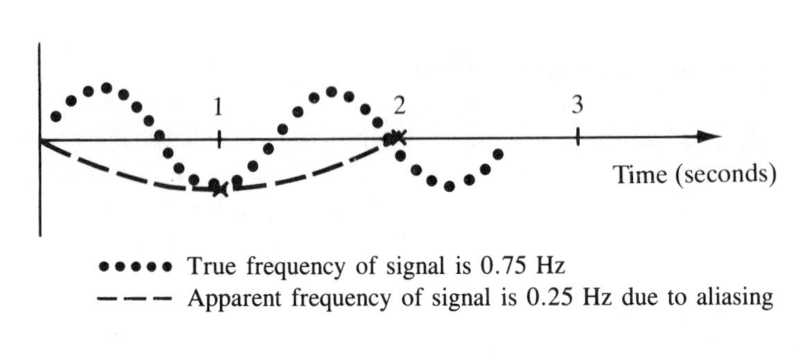

(b) Discrete sampled signal

Figure 5.20 ***Conversion of continuous analog signal to discrete sampled signal***

per second, reconstruction of the samples matches the original analog signal very well. If the rate of sampling were decreased, the fit between the reconstructed samples and the original signal would be less accurate. If the rate of sampling was very much less than the frequency of the raw analog signal, such as one sample per second, only the samples marked 'X' in Figure 5.20 would be obtained. Fitting a line through these Xs incorrectly estimates a signal whose frequency is approximately 0.25 cycles per second. This phenomenon, whereby the process of sampling transmutes a high-frequency signal into a lower-frequency one, is known as *aliasing*. To avoid aliasing, it is theoretically necessary for the sampling rate to be at least twice the highest frequency in the analog signal sampled. In practice, sampling rates of between five and ten times the highest frequency signal are normally chosen so that the discrete sampled signal is a close approximation to the original analogue signal in amplitude as well as frequency.

Problems can arise in sampling when the raw analog signal is corrupted by high frequency noise of unknown characteristics. It would be normal practice to choose the sampling interval as, say, a ten-times multiple of the frequency of the measurement component in the raw signal. If such a sampling interval is chosen, aliasing can in certain circumstances transmute high-frequency noise components into the same frequency range as the measurement component in the signal, thus giving erroneous results. This is one of the circumstances mentioned earlier, where prior analog signal conditioning in the form of a low-pass filter must be carried out before processing the signal digitally.

One further factor which affects the quality of a signal when it is converted from analog to digital form is *quantization*. Quantization describes the procedure whereby the continuous analog signal is converted into a number of discrete levels. At any particular value of the analog signal, the digital representation is either the discrete level immediately above this value or the discrete level immediately below this value. If the difference between two successive discrete levels is represented by the parameter Q, then the maximum error in each digital sample of the raw analog signal is $\pm Q/2$. This error is known as the quantization error and is clearly proportional to the resolution of the analog-to-digital converter, i.e. to the number of bits used to represent the samples in digital form.

Once a satisfactory digital representation in discrete form of an analog signal has been obtained, the procedures of signal amplification, signal attenuation and bias removal become trivial. For signal amplification and attenuation, all samples have to be multiplied or divided by a fixed constant. Bias removal involves simply adding or subtracting a fixed constant from each sample of the signal.

Signal linearization requires *a priori* knowledge of the type of non-linearity involved, in the form of a mathematical equation which expresses the relationship between the output measurements from an instrument and the value of the physical quantity being measured. This can be obtained either theoretically through knowledge of the physical laws governing the system, or empirically using input–output data obtained from the measurement system under controlled conditions. Once this relationship has been obtained, it is used to calculate the value of the measured physical quantity corresponding to each discrete sample of the measurement signal. While the amount of computation involved in this is greater than for the trivial cases of signal amplification, etc., already mentioned, the computational burden is still relatively small in most measurement situations.

Digital signal processing can also perform all of the filtering functions mentioned earlier in respect of analog filters, i.e. low-pass, high-pass, band-pass and band-stop. However, the design of digital filters requires a level of theoretical knowledge, including the use of z-transform theory, which is outside the range of this book. The reader interested in digital filter design is therefore referred elsewhere (Huelsman, 1970; Lynn, 1973).

5.8 Signal transmission

There is a necessity in many measurement systems to transmit measurement signals over quite large distances from the point of measurement to the place where the signals are recorded and/or used in a process-control system. This creates several problems for which a solution must be found. Of the difficulties associated with long-distance signal transmission, the most serious is contamination of the measurement signal by noise. Many sources of noise exist in industrial environments, such as radiated electromagnetic fields from electrical machinery and power cables, induced fields through wiring loops and voltage spikes on the alternating current (a.c.) power supply.

5.8.1 *Signal amplification*

The output signal levels from many types of measurement transducer are relatively low, and amplification of the signal prior to transmission is essential if a reasonable signal-to-noise ratio is to be obtained after transmission. Amplification at the input to the transmission system is also required to compensate for the attenuation of the signal which results from the resistance of the signal wires. The means of amplifying signals has already been discussed in section 5.1.

5.8.2 *Shielding*

Shielding consists of surrounding the signal wires in a cable with a braided metal shield which is connected to earth. This provides a high degree of noise protection, especially against capacitive-induced noise due to the proximity of signal wires to high-current power conductors.

5.8.3 *Current loop transmission*

The signal-attenuation effect of conductor resistances can be minimized if varying voltage signals are transmitted as varying current signals. This requires a voltage-to-current converter of the form shown in Figure 5.21, which is commonly known as a 4–20 mA current loop interface. Two voltage-controlled current sources are used, one (I_1) providing a constant 4 mA output which is used as the power supply current, and the other (I_2) providing a variable 0–16 mA output which is proportional to the input voltage level. The net output current therefore varies between 4 and 20 mA. This is a very commonly used means of connecting remote instruments to a central control room.

5.8.4 *Digital transmission (voltage-to-frequency conversion)*

Even better immunity to noise can be obtained in signal transmission if the signal is transmitted in a digital format. This is achieved by applying the input analog voltage signal to the input of a voltage-to-frequency converter circuit which converts the voltage variations into corresponding frequency variations. Such frequency variations can then be readily transmitted in a digital format. After transmission, reconversion to an analog voltage signal is possible using a frequency-to-voltage converter.

5.8.5 *Fiber-optic transmission*

Fiber-optic signal transmission involves transforming electrical signals into a modulated light wave which is transmitted along a fiber-optic cable. Then, at the receiving end of the cable, the light is transformed back into electrical form.

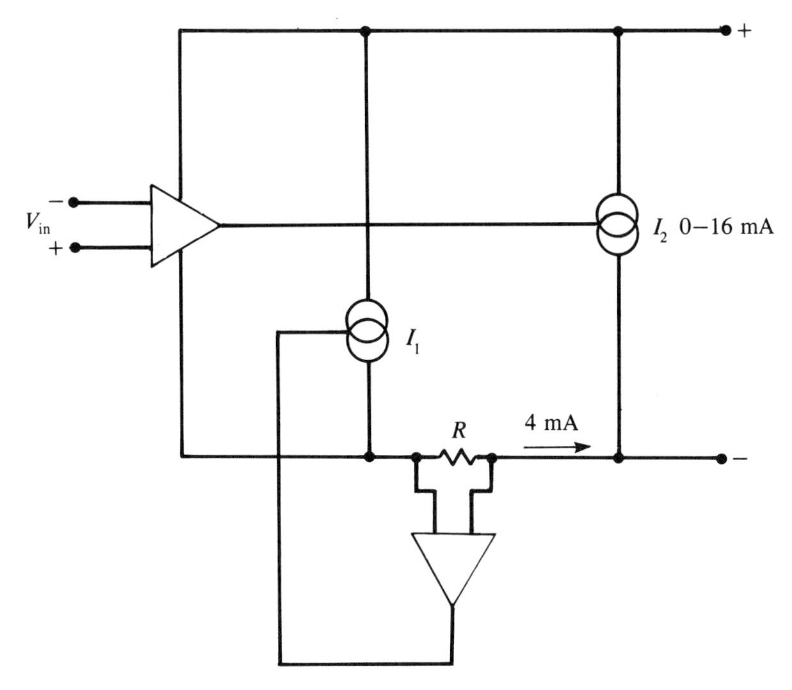

Figure 5.21 *Loop transmitter for 4–20 mA current*

Light has a number of advantages over electricity as a medium for transmitting information: it is immune to corruption by neighbouring electromagnetic fields, the attenuation over a given transmitted distance is much less and it is also intrinsically safe. However, there is an associated cost penalty because of the higher cost of a fiber-optic system compared with the cost of metal conductors. The primary reason for this penalty is the high cost of the terminating transducers which perform the signal conversion function at each end of the cable.

The light-transmitting cable contains at least one, but more often a bundle, of glass or plastic fibers. This is terminated at each end by a transducer, as shown in Figure 5.22. At the input end, the transducer converts the signal from the electrical form, in which most signals originate, into light. At the output end, the transducer converts the transmitted light back into an electrical form suitable for use by data recording, manipulation and display systems. These two transducers are often known as the transmitter and receiver respectively.

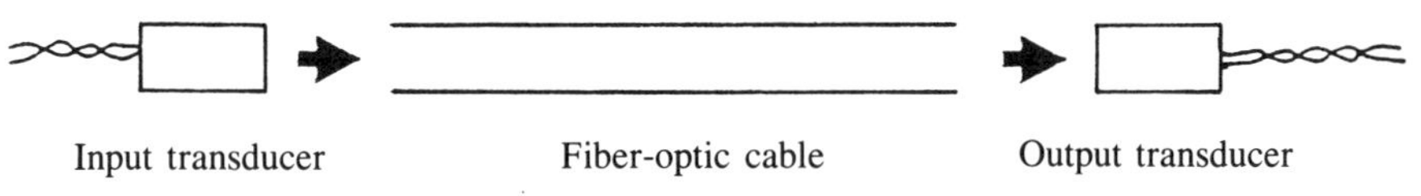

Figure 5.22 *Fiber-optic signal transmission*

Signals are normally transmitted along a fiber-optic cable in digital format, although analog transmission is sometimes used. If there is a requirement to transmit more than one signal, it is more economic to multiplex the signals onto a single cable rather than transmit the signals separately on multiple cables. *Time division multiplexing* involves switching the analog signals in turn, in a synchronized sequential manner, into an analog-to-digital converter which outputs to the transmission line. At the other end of the transmission line, a digital-to-analog converter transforms the digital signal back to analog form and it is then switched in turn to separate analog signal lines.

References and further reading

Blinchikoff, H.J., *Filtering in the Time and Frequency Domains*, 1976. Wiley, New York.
Hilburn, J.L. and Johnson, D.E., *Manual of Active Filter Design*, 1973. McGraw-Hill, New York.
Huelsman, L.P., *Active Filters: Lumped, Distributed, Integrated, Digital and Parametric*, 1970. McGraw-Hill, New York.
Lynn, P.A., *The Analysis and Processing of Signals*, 1973. Macmillan, London.
Skilling, H.H., *Electrical Engineering Circuits*, 1967. Wiley, New York.
Stephenson, F.W., *RC Active Filter Design Handbook*, 1985. Wiley, New York.
Williams, E., *Electric Filter Circuits*, 1963. Pitman, London.

CHAPTER 6

Bridge circuits

Bridge circuits are used very commonly as a variable conversion element in measurement systems. They produce an output in the form of a voltage level which changes as the measured physical quantity changes in value. They provide an accurate method of measuring resistance, inductance and capacitance values, and enable very small changes in these quantities about a nominal value to be detected. They are of immense importance in measurement system technology because so many transducers measuring physical quantities have an output which is expressed as a change in resistance, inductance or capacitance. The displacement-measuring strain gauge, which has a varying resistance output, is but one example of this class of transducers. Excitation of the bridge is normally by a d.c. voltage for resistance measurement and by an a.c. voltage for inductance or capacitance measurement. Both null and deflection types of bridge exist, and, in a similar manner to instruments in general, null types are mainly employed for calibration purposes and deflection types used within closed-loop automatic control schemes.

6.1 Null-type, d.c. bridge (Wheatstone bridge)

A null-type bridge with d.c. excitation, commonly known as a Wheatstone bridge, has the form shown in Figure 6.1. The four arms of the bridge consist of the unknown resistance R_u, two equal value resistors R_2 and R_3 and a variable resistor R_v which is usually a decade resistance box. A d.c. voltage V_i is applied across the points AC and the resistance R_v is varied until the voltage measured across points BD is zero, at which time the bridge is said to be balanced. This null point is usually detected with a high-sensitivity galvanometer.

To analyze the Wheatstone bridge, define the current flowing in each arm to be $I_1 \ldots I_4$ as shown in Figure 6.1. Normally, if a high-impedance voltage measuring instrument is used, the current I_m drawn by the measuring instrument will be very small and can be approximated to zero. If this assumption is made, then, for $I_m = 0$, $I_1 = I_3$ and $I_2 = I_4$.

Looking at path ADC, we have a voltage V_i applied across a resistance $R_u + R_3$ and by Ohm's law:

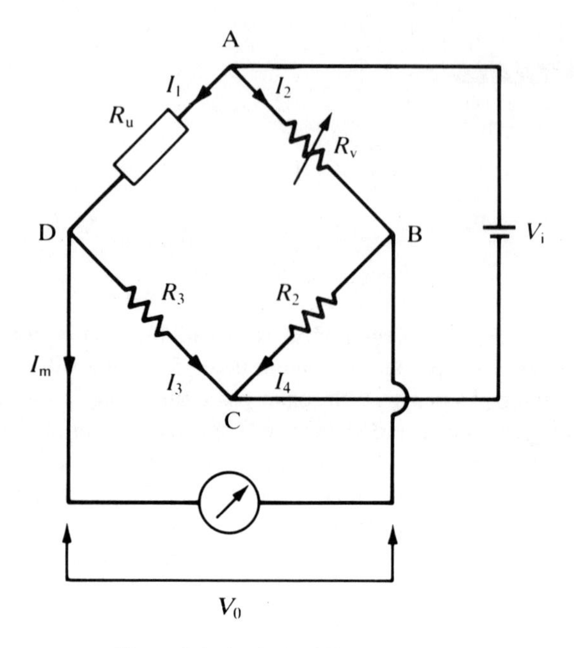

Figure 6.1 *Analysis of Wheatstone bridge*

$$I_1 = \frac{V_i}{R_u + R_3} \tag{6.1}$$

Similarly for path ABC:

$$I_2 = \frac{V_i}{R_v + R_2} \tag{6.2}$$

Now we can calculate the voltage drop across AD and AB:

$$V_{AD} = I_1 \cdot R_u = \frac{V_i \cdot R_u}{R_u + R_3}$$

$$V_{AB} = I_2 \cdot R_v = \frac{V_i \cdot R_v}{R_v + R_2}$$

By the principle of superposition,

$$V_o = V_{BD} = V_{BA} + V_{AD} = -V_{AB} + V_{AD}$$

Thus:

$$V_o = -\frac{V_i \cdot R_v}{R_v + R_2} + \frac{V_i \cdot R_u}{R_u + R_3} \tag{6.3}$$

At balance, $V_o = 0$, and so from equation (6.3):

$$-\frac{V_i \cdot R_v}{R_v + R_2} + \frac{V_i \cdot R_u}{R_u + R_3} = 0$$

and therefore

$$\frac{R_u}{R_u + R_3} = \frac{R_v}{R_v + R_2}$$

Inverting both sides:

$$\frac{R_u + R_3}{R_u} = \frac{R_v + R_2}{R_v}$$

i.e.

$$\frac{R_3}{R_u} = \frac{R_2}{R_v} \quad \text{or} \quad R_u = \frac{R_3 \cdot R_v}{R_2} \tag{6.4}$$

Thus, if $R_2 = R_3$, then $R_u = R_v$.

The value of R_v is accurately known because it is derived from a calibrated decade resistance box and this means that R_u is also accurately known.

6.2 Deflection-type d.c. bridge

A deflection-type bridge with d.c. excitation is shown in Figure 6.2. This differs from the Wheatstone bridge mainly in that the variable resistance R_v is replaced by a fixed resistance R_1 of the same value as the nominal value of the unknown resistance R_u. As the resistance R_u changes, so the output voltage V_o varies, and this relationship between V_o and R_u must be calculated.

This relationship is simplified if we again assume that a high-impedance voltage measuring instrument is used so that the current I_m in the instrument can be approximated to zero. (The case when this assumption does not hold is covered in more advanced texts, e.g. Morris, 1993.) The analysis is then exactly the same as for the preceding example of the Wheatstone bridge, except that R_v is replaced by R_1. Thus, from equation (6.3), we have:

$$V_o = V_i\left(\frac{R_u}{R_u + R_3} - \frac{R_1}{R_1 + R_2}\right) \tag{6.5}$$

When R_u is at its nominal value, i.e. for $R_u = R_1$, it is clear that $V_o = 0$, since $R_2 = R_3$. For other values of R_u, V_o has negative or positive values which vary in a non-linear way with R_u.

EXAMPLE 6.1

A certain type of pressure transducer, designed to measure pressures in the range

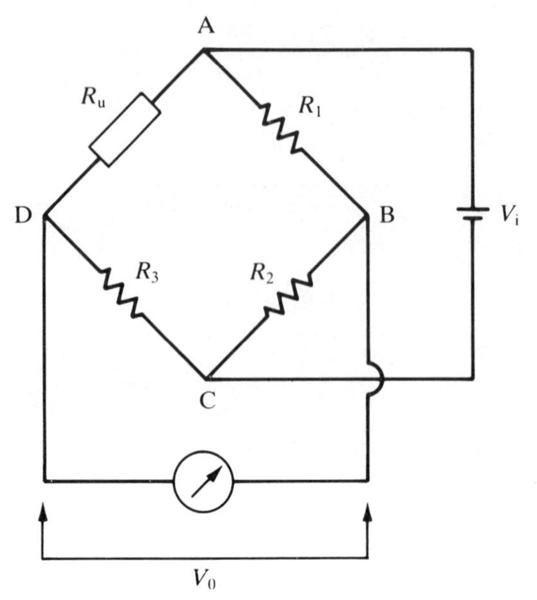

Figure 6.2 *Deflection-type d.c. bridge*

0–10 bar, consists of a diaphragm with a strain gauge cemented to it to detect diaphragm deflections. The strain gauge has a nominal resistance of 120 Ω and forms one arm of a Wheatstone bridge circuit, with the other three arms each having a resistance of 120 Ω. The bridge output is measured by an instrument whose input impedance can be assumed infinite. If, in order to limit heating effects, the maximum permissible gauge current is 30 mA, calculate the maximum permissible bridge excitation voltage. If the sensitivity of the strain gauge is 338 mΩ/bar and the maximum bridge excitation voltage is used, calculate the bridge output voltage when measuring a pressure of 10 bar.

SOLUTION

This is the type of bridge circuit shown in Figure 6.2 in which the components have the following values:

$$R_1 = R_2 = R_3 = 120\,\Omega$$

Defining I_1 to be the current in path ADC of the bridge, we can write:

$$V_i = I_1(R_u + R_3)$$

At balance, $R_u = 120\,\Omega$ and the maximum value allowable for I_1 is 0.03 A. Hence:

$$V_i = 0.03(120 + 120) = 7.2\,V$$

Thus the maximum bridge excitation voltage allowable is 7.2 volts.

For a pressure of 10 bar applied, the resistance change is 3.38 Ω, i.e. R_u is then equal to 123.38 Ω. Applying equation (6.5), we can write:

$$V_o = V_i \left[\frac{R_u}{R_u + R_3} - \frac{R_1}{R_1 + R_2} \right] = 7.2 \left[\frac{123.38}{243.38} - \frac{120}{240} \right] = 50\,\text{mV}$$

Thus, if the maximum permissible bridge excitation voltage is used, the output voltage is 50 mV when a pressure of 10 bar is measured.

The non-linear relationship between output reading and measured quantity expressed by equation (6.5) is inconvenient and does not conform with our normal requirement for a linear input–output relationship. The method of coping with this non-linearity varies according to the form of primary transducer involved in the measurement system.

One special case is where the change in the unknown resistance R_u is typically small compared with the nominal value of R_u. If we calculate the new voltage V_o' when the resistance R_u in equation (6.5) changes by an amount δR_u, we have:

$$V_o' = V_i \left[\frac{R_u + \delta R_u}{R_u + \delta R_u + R_3} - \frac{R_1}{R_1 + R_2} \right] \tag{6.6}$$

The change of voltage output is therefore given by:

$$\delta V_o = V_o' - V_o = \frac{V_i \, \delta R_u}{R_u + \delta R_u + R_3}$$

If $\delta R_u \ll R_u$, then the following linear relationship is obtained:

$$\frac{\delta V_o}{\delta R_u} \approx \frac{V_i}{R_u + R_3} \tag{6.7}$$

This expression describes the measurement sensitivity of the bridge. Such an approximation to make the relationship linear is valid for transducers such as strain gauges where the typical changes of resistance with strain are very small compared with the nominal gauge resistance.

However, many instruments which are inherently linear themselves at least over a limited measurement range, such as resistance thermometers, exhibit large changes in output as the input quantity changes, and the approximation of equation (6.7) cannot be applied. In such cases, specific action must be taken to improve linearity in the relationship between the bridge output voltage and the measured quantity. One common solution to this problem is to make the values of the resistances R_2 and R_3 at least ten times that of R_1. The effect of this is best observed by looking at a numerical example.

Consider a platinum resistance thermometer with a range of 0–50 °C, whose resistance at 0 °C is 500 Ω and whose resistance varies with temperature at the rate of 4 Ω/°C. Over this range of measurement, the output characteristic of the thermometer itself is nearly perfectly linear. (NB The subject of resistance thermometers is discussed further in Chapter 9.)

Taking first the case where $R_1 = R_2 = R_3 = 500\,\Omega$ and $V_i = 10\,\text{V}$, and applying equation (6.5):

at 0 °C; $V_o = 0$

$$\text{at } 25\,°\text{C};\quad R_u = 600\,\Omega;\quad V_o = 10\left[\frac{600}{1100} - \frac{500}{1000}\right] = 0.455\,\text{V}$$

$$\text{at } 50\,°\text{C};\quad R_u = 700\,\Omega;\quad V_o = 10\left[\frac{700}{1200} - \frac{500}{1000}\right] = 0.833\,\text{V}$$

This relationship between V_o and R_u is plotted as curve A in Figure 6.3 and the non-linearity is apparent. Inspection of the changes in the output voltage V_o for equal steps of temperature change also clearly demonstrates the non-linearity:

1 For the temperature change from 0 to 25 °C, the change in V_o is (0.455 − 0) = 0.455 V
2 For the temperature change from 25 to 50 °C, the change in V_o is (0.833 − 0.455) = 0.378 V

If the relationship was linear, the change in V_o for the 25–50 °C temperature step would also be 0.455 V, giving a value for V_o of 0.910 V at 50 °C.

Now take the case where $R_1 = 500\,\Omega$ but $R_2 = R_3 = 5000\,\Omega$ and apply equation (6.5) with $V_i = 26.1\,V$ (the reason for changing V_i will be explained later):

at 0 °C; $V_o = 0$

$$\text{at } 25\,°\text{C};\quad R_u = 600\,\Omega;\quad V_o = 26.1\left[\frac{600}{5600} - \frac{500}{5500}\right] = 0.424\,\text{V}$$

$$\text{at } 50\,°\text{C};\quad R_u = 700\,\Omega;\quad V_o = 26.1\left[\frac{700}{5700} - \frac{500}{5500}\right] = 0.833\,\text{V}$$

This relationship is shown as curve B in Figure 6.3 and a considerable improvement in linearity is achieved. This is more apparent if the differences in values for V_o over the two temperature steps are inspected:

1 From 0 to 25 °C, the change in V_o is 0.424 V.
2 From 25 to 50 °C, the change in V_o is 0.409 V.

The changes in V_o over the two temperature steps are much closer to being equal than before, demonstrating the improvement in linearity.

In increasing the values of R_2 and R_3, it was also necessary to increase the excitation voltage from 10 V to 26.1 V to obtain the same output level at 50 °C. In practical applications, V_i would normally be set at the maximum level consistent with the limitation of the effect of circuit heating in order to maximize the measurement sensitivity ($\delta V_o/\delta R_u$ relationship). It would therefore not be possible to increase V_i

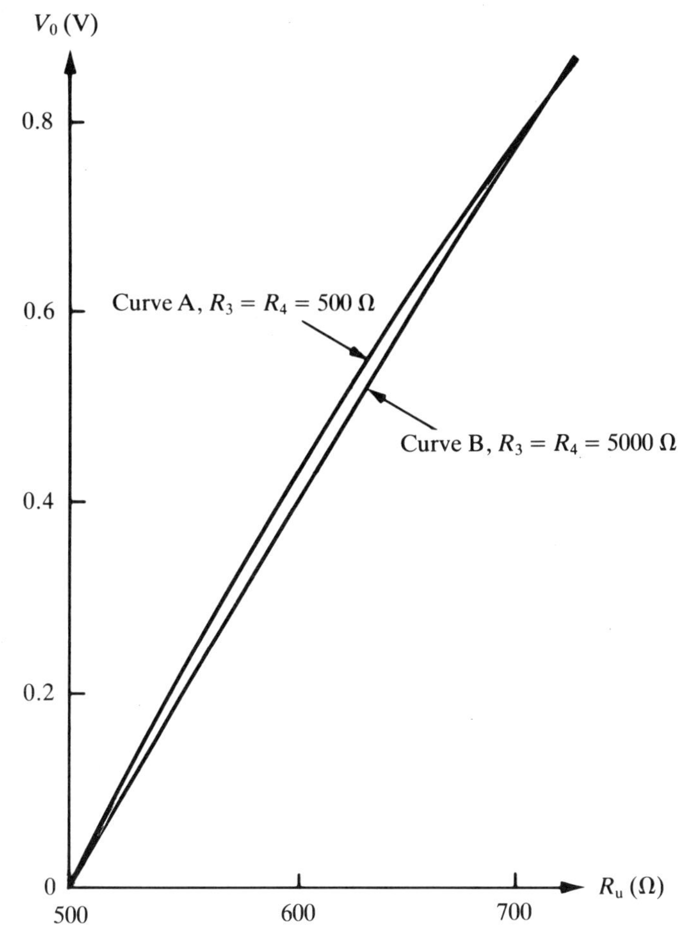

Figure 6.3 *Linearization of bridge circuit characteristic*

further if R_2 and R_3 were increased, and the general effect of such an increase in R_2 and R_3 is thus a decrease in the sensitivity of the measurement system.

The importance of this inherent non-linearity in the bridge output relationship is greatly diminished if the primary transducer and bridge circuit are incorporated as elements within an intelligent instrument. In that case, digital computation is applied to produce an output in terms of the measured quantity which automatically compensates for the non-linearity in the bridge circuit.

6.3 **Error analysis**

In the application of bridge circuits, the contribution of component-value tolerances to total measurement system accuracy limits must be clearly understood. The following analysis applies to a null-type bridge, but similar principles can be applied for a deflection-type bridge.

The maximum measurement error is determined by first finding the value of R_u in equation (6.4) with each parameter in the equation set at that limit of its tolerance which produces the maximum value of R_u. Similarly, the minimum possible value of R_u is calculated, and the required error band is then the span between these maximum and minimum values.

EXAMPLE 6.2

In the Wheatstone bridge circuit of Figure 6.1, R_v is a decade resistance box with a quoted inaccuracy of ±0.2% and $R_2 = R_3 = 500\,\Omega \pm 0.1\%$. If the value of R_v at the null position is 520.4 Ω, determine the error band for R_u expressed as a percentage of its nominal value.

SOLUTION

Applying equation (6.4) with $R_v = 520.4\,\Omega + 0.2\% = 521.44\,\Omega$, $R_3 = 5000\,\Omega + 0.1\% = 5005\,\Omega$, $R_2 = 5000\,\Omega - 0.1\% = 4995\,\Omega$, we get:

$$R_u = \frac{521.44 \times 5005}{4995} = 522.48\,\Omega \quad (= +0.4\%)$$

Applying equation (6.4) with $R_v = 520.4\,\Omega - 0.2\% = 519.36\,\Omega$, $R_3 = 5000\,\Omega - 0.1\% = 4995\,\Omega$, $R_2 = 5000\,\Omega + 0.1\% = 5005\,\Omega$, we get:

$$R_u = \frac{519.36 \times 4995}{5005} = 518.32\,\Omega \quad (= -0.4\%)$$

Thus the error band for R_u is ±0.4%.

The cumulative effect of errors in individual bridge circuit components is clearly seen. Although the maximum error in any one component is ±0.2%, the possible error in the measured value of R_u is ±0.4%. Such a magnitude of error is often not acceptable, and special measures are taken to overcome the introduction of error by component-value tolerances. One such practical measure is the introduction of apex balancing. This is one of many methods of bridge balancing which all produce a similar result.

6.3.1 Apex balancing

One form of apex balancing consists of placing an additional variable resistor R_5 at the junction C between the resistances R_2 and R_3, and applying the excitation voltage V_i to the wiper of this variable resistance, as shown in Figure 6.4.

For calibration purposes, R_u and R_v are replaced by two equal resistances whose values are accurately known, and R_5 is adjusted until the output voltage V_o is zero. At this point, if the portions of resistance on either side of the wiper on R_5 are R_6 and R_7 (such that $R_5 = R_6 + R_7$), we can write:

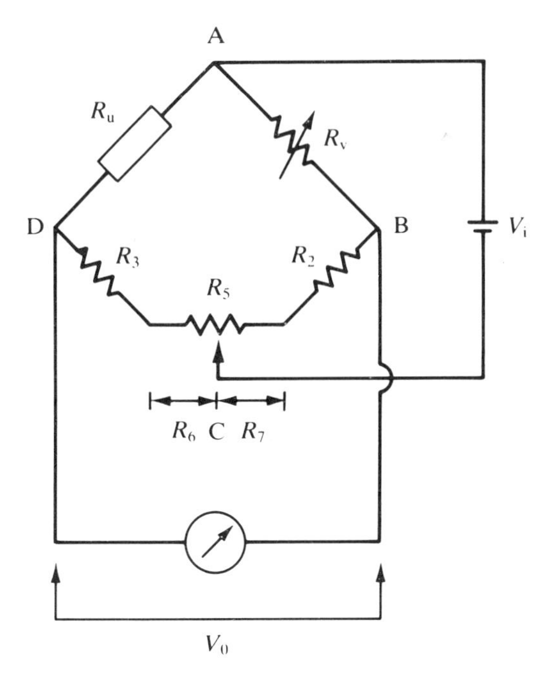

Figure 6.4 *Apex balancing*

$$R_3 + R_6 = R_2 + R_7$$

We have thus eliminated any source of error due to the tolerance in the value of R_2 and R_3, and the error in the measured value of R_u depends only on the accuracy of one component, the decade resistance box R_v.

EXAMPLE 6.3

A potentiometer R_5 is put into the apex of the bridge shown in Figure 6.4 to balance the circuit. The bridge components have the following values: R_u = 500 Ω, R_v = 500 Ω, R_2 = 515 Ω, R_3 = 480 Ω and R_5 = 100 Ω. Determine the required value of the resistances R_6 and R_7 of the parts of the potentiometer track either side of the slider in order to balance the bridge and compensate for the unequal values of R_2 and R_3.

SOLUTION

For balance, $R_2 + R_7 = R_3 + R_6$; hence, $515 + R_7 = 480 + R_6$. Also, because R_6 and R_7 are the two parts of the potentiometer track R_5 whose resistance is 100 Ω: $R_6 + R_7 = 100$. Thus

$$515 + R_7 = 480 + (100 - R_7)$$

i.e. $2R_7 = 580 - 515 = 65$. Thus, $R_7 = 32.5\,\Omega$. Hence

$$R_6 = 100 - 32.5 = 67.5\,\Omega.$$

6.4 A.C. bridges

Bridges with a.c. excitation are used to measure unknown impedances. As for d.c. bridges, both null and deflection types exist, with null types being generally reserved for calibration duties.

6.4.1 Null-type impedance bridge

A typical null-type impedance bridge is shown in Figure 6.5. The null point can be conveniently detected by monitoring the output with a pair of headphones connected via an operational amplifier across the points BD. This is a much cheaper method of null detection than the application of an expensive galvanometer which is required for a d.c. Wheatstone bridge.

Referring to Figure 6.5, at the null point:

$$I_1 \cdot R_1 = I_2 \cdot R_2; \quad I_1 \cdot Z_u = I_2 \cdot Z_v$$

Thus:

$$Z_u = \frac{Z_v \cdot R_1}{R_2} \tag{6.8}$$

If Z_u is capacitive, i.e. $Z_u = 1/j\omega C_u$, then Z_v must consist of a variable capacitance box, which is readily available. If Z_u is inductive, then $Z_u = R_u + j\omega L_u$.

Notice that the expression for Z_u now has a resistive term in it because it is impossible to realize a pure inductor. An inductor coil always has a resistive

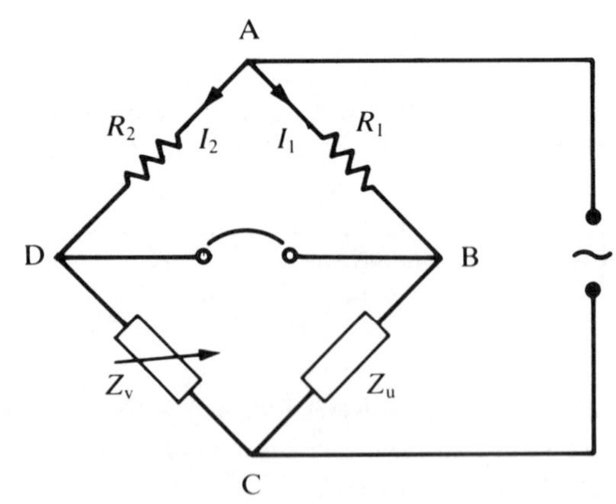

Figure 6.5 *Null-type impedance bridge*

component, though this is made as small as possible by designing the coil to have a high Q factor (Q factor is the ratio inductance/resistance).

Therefore, Z_v must consist of a variable-resistance box and a variable-inductance box. However, the latter is not readily available because it is difficult and hence expensive to manufacture a set of fixed value inductors to make up a variable-inductance box.

For this reason, an alternative kind of null-type bridge circuit, known as the Maxwell bridge, is commonly used to measure unknown inductances.

6.4.2 **Maxwell bridge**

A Maxwell bridge is shown in Figure 6.6. The requirement for a variable inductance box is avoided by introducing instead a second variable resistance. The circuit requires one standard fixed-value capacitor (C), two variable resistance boxes (R_1 and R_2) and one fixed-value resistor (R_3), all of which are components which are readily available and relatively inexpensive.

Referring to Figure 6.6, we have when the bridge is balanced:

$$I_1 Z_{AD} = I_2 Z_{AB}; \quad I_1 Z_{DC} = I_2 Z_{BC}$$

Thus:

$$\frac{Z_{BC}}{Z_{AB}} = \frac{Z_{DC}}{Z_{AD}} \quad \text{or} \quad Z_{BC} = \frac{Z_{DC} Z_{AB}}{Z_{AD}} \tag{6.9}$$

The quantities in equation (6.9) have the following values:

$$\frac{1}{Z_{AD}} = \frac{1}{R_1} + j\omega C \quad \text{or} \quad Z_{AD} = \frac{R_1}{1 + j\omega R_1 C}$$

$$Z_{AB} = R_3; \quad Z_{BC} = R_u + j\omega L_u; \quad Z_{DC} = R_2$$

Substituting the values into equation (6.9):

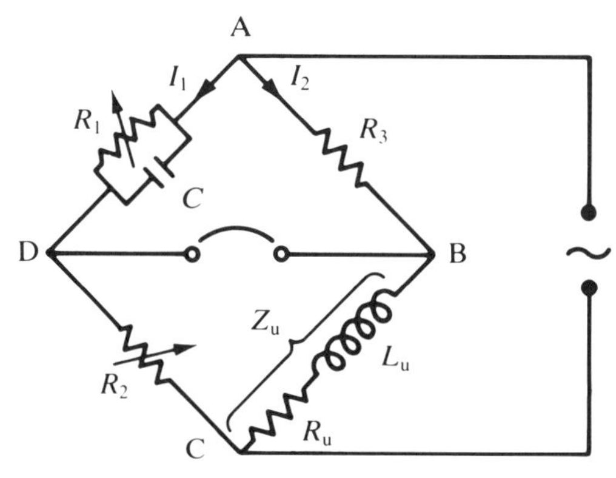

Figure 6.6 *Maxwell bridge*

$$R_u + j\omega L_u = \frac{R_2R_3(1 + j\omega CR_1)}{R_1}$$

Taking real and imaginary parts:

$$R_u = \frac{R_2R_3}{R_1}; \quad L_u = R_2R_3C \tag{6.10}$$

Expression (6.10) can be used to calculate the quality factor (Q value) of the coil thus:

$$Q = \frac{\omega L_u}{R_u} = \frac{\omega R_2R_3CR_1}{R_2R_3} = \omega CR_1$$

If a constant frequency ω is used:

$$Q \propto R_1$$

Thus, the Maxwell bridge can be used to measure the Q value of a coil directly using this relationship.

EXAMPLE 6.4

In the Maxwell bridge shown in Figure 6.6, let the fixed-value bridge components have the following values: $R_3 = 5\,\Omega$; $C = 1$ mF. Calculate the value of the unknown impedance (L_u, R_u) if $R_1 = 159\,\Omega$ and $R2 = 10\,\Omega$ at balance.

SOLUTION

Substituting values into the relations developed in equation (6.10) above:

$$R_u = \frac{R_2R_3}{R_1} = \frac{10 \times 5}{159} = 0.3145\,\Omega$$

$$L_u = R_2R_3C = \frac{10 \times 5}{10^3} = 50\,\text{mH}$$

EXAMPLE 6.5

Calculate the Q factor for the unknown impedance in example 6.5 above at a supply frequency of 50 Hz.

SOLUTION

$$Q = \frac{\omega L_u}{R_u} = \frac{2\pi 50(0.05)}{0.3145} = 49.9$$

6.4.3 *Deflection-type a.c. bridge*

A common deflection-type of a.c. bridge circuit is shown in Figure 6.7. For

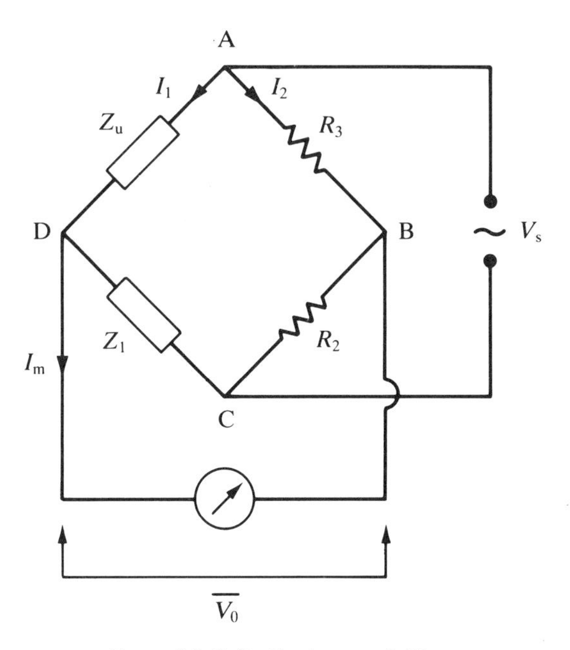

Figure 6.7 *Deflection-type a.c. bridge*

capacitance measurement:

$$Z_u = \frac{1}{j\omega C_u}; \quad Z_1 = \frac{1}{j\omega C_1}$$

For inductance measurement, assuming that the resistive component of the inductor is small and approximates to zero:

$$Z_u = j\omega L_u; \quad Z_1 = j\omega L_1$$

Analysis of the circuit to find the relationship between V_o and Z_u is greatly simplified if one assumes that I_m is very small. For $I_m = 0$, currents in the two branches of the bridge, as defined in Figure 6.7, are given by:

$$I_1 = \frac{V_s}{Z_1 + Z_u}; \quad I_2 = \frac{V_s}{R_2 + R_3}$$

Also $V_{AD} = I_1 \cdot Z_u$ and $V_{AB} = I_2 \cdot R_3$.
Then:

$$V_o = V_{BD} = V_{AD} - V_{AB} = V_s \left[\frac{Z_u}{Z_1 + Z_u} - \frac{R_3}{R_2 + R_3} \right]$$

Thus for capacitances:

$$V_o = V_s\left[\frac{1/C_u}{1/C_1 + 1/C_u} - \frac{R_3}{R_2 + R_3}\right] = V_s\left[\frac{C_1}{C_1 + C_u} - \frac{R_3}{R_2 + R_3}\right] \quad (6.11)$$

and for inductances:

$$V_o = V_s\left[\frac{L_u}{L_1 + L_u} - \frac{R_3}{R_2 + R_3}\right] \quad (6.12)$$

The relationship shown in equation (6.12) is in practice only approximate since inductive impedances are never pure inductances as assumed but always contain a finite resistance (i.e.$Z_u = j\omega L_u + R$). However, the approximation is valid in many circumstances.

EXAMPLE 6.6

A deflection bridge as shown in Figure 6.7 is used to measure an unknown capacitance, C_u. The components in the bridge have the following values: $V_s =$ 20 Vrms,[1] $C_1 = 100\,\mu\text{F}$, $R_2 = 60\,\Omega$, $R_3 = 40\,\Omega$. If $C_u = 100\,\mu\text{F}$, calculate the output voltage V_o.

SOLUTION

From equation (6.11)

$$V_o = V_s\left[\frac{C_1}{C_1 + C_u} - \frac{R_3}{R_2 + R_3}\right] = 20(0.5 - 0.4) = 2\,\text{Vrms}$$

EXAMPLE 6.7

An unknown inductance L_u is measured using a deflection-type of bridge as shown in Figure 6.7. The components in the bridge have the following values: V_s = 10 Vrms, $L_1 = 20\,\text{mH}$, $R_2 = 100\,\Omega$, $R_3 = 100\,\Omega$. If the output voltage V_O is 1 Vrms, calculate the value of L_u.

SOLUTION

From equation (6.12)

$$\frac{L_u}{L_1 + L_u} = \frac{V_o}{V_s} + \frac{R_3}{R_2 + R_3} = 0.1 + 0.5 = 0.6$$

Thus $L_u = 0.6(L_1 + L_u)$ and therefore $0.4L_u = 0.6L_1$. Hence:

$$L_u = \frac{0.6 \times 0.02}{0.4} = 30\,\text{mH}$$

[1] This means a root-mean square voltage magnitude of 20 volts.

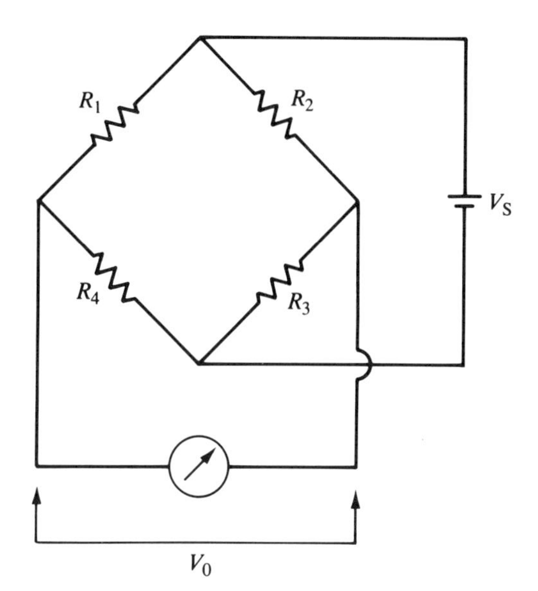

Figure 6.8 *Wheatstone bridge*

6.5 Self-assessment questions

6.1 (a) A Wheatstone bridge circuit is shown in Figure 6.8. Derive an expression for the output voltage V_0 in terms of R_1, R_2, R_3, R_4 and V_s, assuming that the impedance of the voltage-measuring instrument is infinite.
(b) If the elements have the following values: $R_1 = 110\,\Omega$, $R_2 = 100\,\Omega$, $R_3 = 1000\,\Omega$, $R_4 = 1000\,\Omega$, and $V_s = 10\,\text{V}$, calculate the output voltage V_o.

6.2 In the Wheatstone bridge shown in Figure 6.8, the resistive components have the following nominal values: $R_1 = 3\,\text{k}\Omega$, $R_2 = 6\,\text{k}\Omega$, $R_3 = 8\text{k}\Omega$ and $R_4 = 4\text{k}\Omega$. The actual value of each resistance is related to the nominal value according to: $R_{actual} = R_{nominal} + \delta R$ where δR has the following values: $\delta R_1 = 30\,\Omega$, $\delta R_2 = -20\Omega$, $\delta R_3 = 40\,\Omega$ and $\delta R_4 = -50\,\Omega$. Calculate the open circuit bridge output voltage if the bridge supply voltage V_s is 50 V.

6.3 (a) Figure 6.2 shows a d.c. bridge circuit designed to measure the value of an unknown resistance R_u. The impedance of the instrument measuring the output voltage V_o is such that the current in it is negligible. Derive an expression for the output voltage V_o in terms of the input voltage V_i and the four resistances in the bridge, R_u, R_1, R_2, and R_3.
(b) Suppose that the unknown resistance R_u is a resistance thermometer whose resistance at 100 °C is 500 Ω and whose resistance varies with temperature at the rate of 0.5 Ω/°C for small temperature changes around 100 °C. Calculate the sensitivity of the total measurement system for small changes in temperature around 100 °C, given the following resistance and voltage values measured at

15 °C by instruments calibrated at 15 °C: $R_1 = 500\ \Omega$; $R_2 = R_3 = 5000\ \Omega$ and $V_i = 10$ V.

(c) If the resistance thermometer is measuring a fluid whose true temperature is 104 °C, calculate the error in the indicated temperature if the ambient temperature around the bridge circuit is 20 °C instead of the calibration temperature of 15 °C, given the following additional information: voltage-measuring instrument zero drift coefficient = +1.3 mV/°C; voltage-measuring instrument sensitivity drift coefficient is zero; resistances R_1, R_2 and R_3 have a positive temperature coefficient of +0.2% of nominal value/°C; voltage source V_i is unaffected by temperature changes.

6.4 Four strain gauges of resistance 120 Ω each are arranged into a Wheatstone bridge configuration such that each of the four arms in the bridge has one strain gauge in it. The maximum permissible current in each strain gauge is 100 mA. What is the maximum bridge supply voltage allowable, and what power is dissipated in each strain gauge with that supply voltage?

6.5 (a) Suppose that the parameters shown in Figure 6.2 have the following values: $R_1 = 100\ \Omega$, $R_2 = 100\ \Omega$, $R_3 = 100\ \Omega$ and $V_i = 12$ V. R_u is a resistance thermometer with a resistance of 100 Ω at 100 °C and a temperature coefficient of +0.3 Ω/°C over the temperature range from 50 to 150 °C (i.e. the resistance increases as the temperature goes up). Draw a graph of bridge output voltage V_o for ten degree steps in temperature between 100 and 150 °C (calculating V_o according to equation 6.5).

(b) Draw a graph of V_o for similar temperature values if $R_2 = R_3 = 1000\ \Omega$ and all other components have the same values as given in part (a) above. Notice that the line through the data points is straighter than that drawn in (a) but the output voltage is much less at each temperature point.

6.6 In the d.c. bridge circuit shown in Figure 6.9, the resistive components have the following values: $R_1 = R_2 = 120\ \Omega$, $R_3 = 117\ \Omega$, $R_4 = 123\ \Omega$ and $R_a = R_p = 1000\ \Omega$.

(a) What are the resistance values of the parts of the potentiometer track either side of the slider when the potentiometer is adjusted to balance the bridge?

(b) What then is the effective resistance of each of the two left-hand arms of the bridge when the bridge is balanced?

6.7 A Maxwell bridge, designed to measure the unknown impedance (R_u, L_u) of a coil, is shown in Figure 6.6.

(a) Derive an expression for R_u and L_u under balance conditions.

(b) If the fixed bridge component values are $R_3 = 100\Omega$ and $C = 20\ \mu$F, calculate the value of the unknown impedance if $R_1 = 3183\ \Omega$ and $R_2 = 50\ \Omega$ at balance.

(c) Calculate the Q factor for the coil if the supply frequency is 50 Hz.

6.8 A deflection-type a.c. bridge as shown in Figure 6.7 is used to measure an unknown inductance L_u. The components in the bridge have the following values: $V_s = 30$ Vrms, $L_1 = 80$ mH, $R_2 = 70\ \Omega$, $R_3 = 30\ \Omega$. If $L_u = 50$ mH, calculate the output voltage V_o.

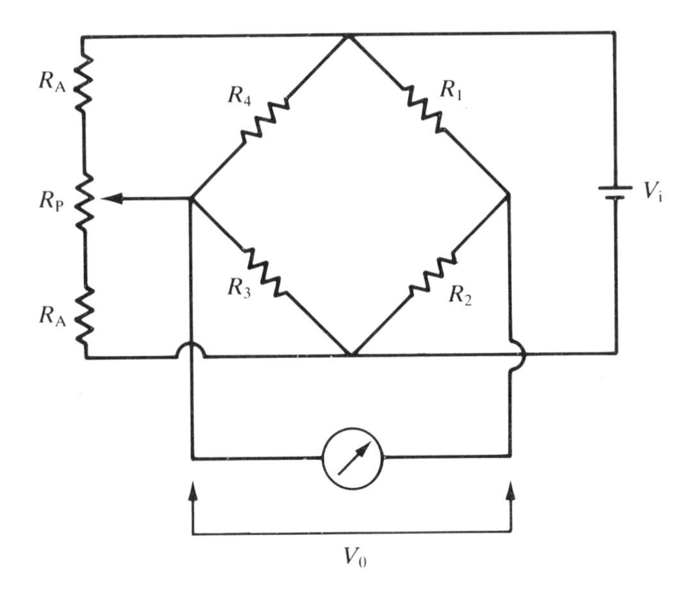

Figure 6.9 *D.C. bridge*

6.9 An unknown capacitance C_u is measured using a deflection-type a.c. bridge as shown in Figure 6.7. The components of the bridge have the following values: $V_s = 10\,\text{Vrms}$, $C_1 = 50\,\mu\text{F}$, $R_2 = 80\,\Omega$ and $R_3 = 20\,\Omega$. If the output voltage is 3 Vrms, calculate the value of C_u.

6.10 A Hays bridge is often used for measuring the inductance of high-Q coils and has the configuration shown in Figure 6.10.
(a) Obtain the bridge balance conditions.
(b) Show that if the Q value of an unknown inductor is high, the expression for the inductance value when the bridge is balanced is independent of frequency.

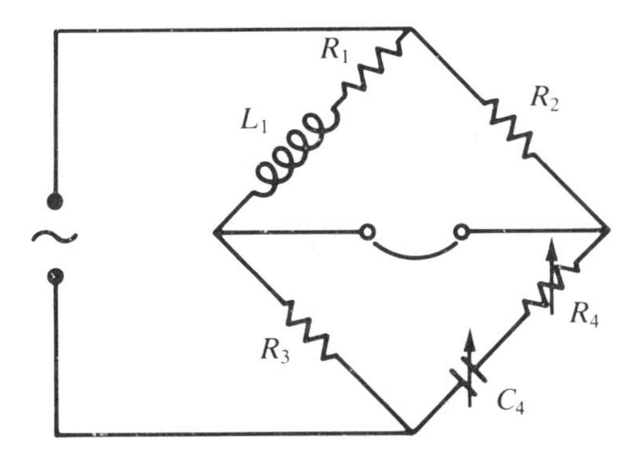

Figure 6.10 *Hays bridge*

(c) If the Q value is high, calculate the value of the inductor if the bridge component values at balance are as follows: $R_2 = R_3 = 1000\,\Omega$; $C = 0.02\,\mu\text{F}$.

References and further reading

Jones, C., *An Introduction to Advanced Electrical Engineering*, 1962. English Universities Press, London.

Morris, A.S., *Principles of Measurement and Instrumentation*, 1993. Prentice-Hall, Hemel Hempstead.

Smith, R.J., *Circuits, Devices and Systems*, 1976. Wiley, New York.

CHAPTER 7

Measurement of electrical signals and related quantities

7.1 Introduction

A large number of measuring instruments and transducers have an output which is in the form of either a change of resistance, capacitance or inductance or alternatively modulation of some other electrical parameter such as the voltage level, current, power level, frequency or phase of a signal. Study of the various techniques available for measuring these quantities is therefore important.

For some purposes, it is only necessary to have a visual indication of the signal level when monitoring the output of a measuring instrument, while for other purposes there is a requirement to record the output signal continuously in a form that can be recorded and saved for future study and use.

The instruments used to give a visual indication of the level of electrical signals are meters, of both analog and digital varieties, and the cathode-ray oscilloscope. These are discussed in sections 7.2 and 7.3 respectively. The later sections in this chapter then discuss measurement of the various other electrical parameters which may be required in a particular measurement system.

7.2 Meters

Meters exist in both digital and analog forms. The digital forms consist of various versions of the digital voltmeter (normally abbreviated to DVM), which are classified according to the method used to convert the analog voltage signal into a corresponding digital output reading. Analog meters can be divided into six sub-classifications, according to their principle of operation. These six types are moving-coil, moving-iron, electrodynamic, induction, clamp-on and electrostatic, respectively.

Analog meters are electromechanical devices driving a pointer against a scale. They are prone to measurement errors from a number of sources, with quoted inaccuracy figures of between ±0.1% and ±3%. Inaccurate scale marking during manufacture, bearing friction, bent pointers and ambient temperature variations all limit measurement accuracy. Further human errors are introduced through mistakes in interpolating between scale markings and through parallax error, which is caused by the reader not taking the measurement from directly above the scale.

Digital meters give a reading in the form of a digital display. There are no problems of parallax and every observer sees the same value. In fact, although usually more expensive than analog meters, digital meters are superior in every other respect and have quoted inaccuracy levels of between ±0.005% and ±2%, depending on the parameter measured and the quality of the instrument. Particular advantages of digital voltmeters include their very high input impedance (typically 10 MΩ compared with 1–20 kΩ for analog meters), the ability to measure signals of frequency up to 1 MHz (though not with the cheapest instruments) and automatic ranging (which prevents overload and reverse polarity connection, etc.).

7.2.1 *Voltage-to-time conversion digital voltmeter*

This is known as a ramp type of DVM. When an unknown voltage signal is applied to the input terminals of the instrument, a negative-slope ramp waveform is generated internally and is compared with the input signal. When the two are equal, a pulse is generated which opens a gate, and at a later point in time a second pulse closes the gate when the negative ramp voltage reaches zero. The length of time between the gate opening and closing is monitored by an electronic counter, which produces a digital display according to the level of the input voltage signal. The lowest inaccuracy figure achievable for this type is about ±0.05%.

7.2.2 *Potentiometric digital voltmeter*

This type of DVM uses the servo principle, where the error between the unknown input voltage level and a reference voltage is applied to a servo-driven potentiometer which adjusts the reference voltage until it balances the unknown voltage. The output reading is produced by a mechanical drum-type digital display driven by the potentiometer. This is a relatively cheap form of DVM.

7.2.3 *Dual-slope integration digital voltmeter*

In this type of DVM, the unknown voltage is applied to an integrator for a fixed time T_1, following which a reference voltage of opposite sign is applied to the integrator, which discharges down to a zero output in an interval T_2 measured by a counter. The output–time relationship for the integrator is shown in Figure 7.1, from which the unknown voltage V_i can be calculated geometrically from the triangle as:

$$V_i = V_{ref} \cdot (T_2/T_1)$$

This is another relatively simple and cheap form of DVM.

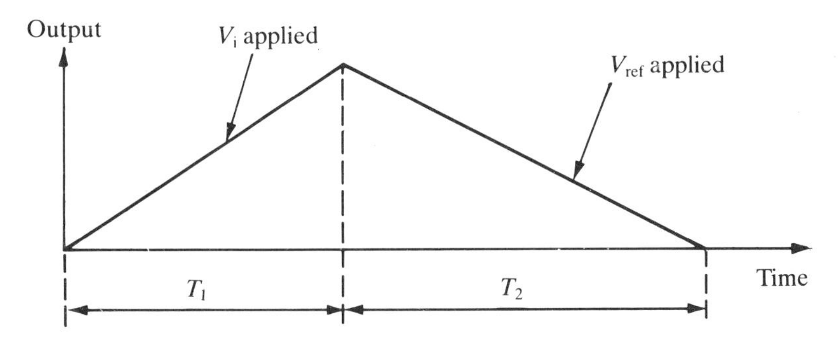

Figure 7.1 *Output–time relationship for the integrator in a dual-slope digital voltmeter (DVM)*

7.2.4 Voltage-to-frequency conversion digital voltmeter

In this form, the unknown voltage signal is fed via a range switch and an amplifier to a converter circuit whose output is in the form of a train of voltage pulses at a frequency proportional to the magnitude of the input signal. The main advantage of this type of DVM is its ability to reject a.c. noise.

7.2.5 Moving-coil meters

A moving-coil meter is an analog instrument which responds only to direct current inputs and is shown schematically in Figure 7.2. It consists of a rectangular coil wound round a soft iron core which is suspended in the field of a permanent magnet. The signal being measured is applied to the coil and this produces a radial magnetic field. Interaction between this induced field and the field produced by the permanent magnet causes a torque, which results in rotation of the coil against the restraining torque of a spring. The amount of rotation of the coil is measured by attaching a pointer to it which moves past a graduated scale.

The theoretical torque T generated is given by:

$$T = BIhwN \tag{7.1}$$

where $\boldsymbol{B}$ is the flux density of the radial field, I is the current in the coil, h is the height of the coil, w is the width of the coil, N is the number of turns in the coil and where the units are as defined in Table 1.2.

If the iron core is cylindrical and the air gap between the coil and pole faces of the permanent magnet is uniform, then the flux density $\boldsymbol{B}$ is constant, and equation (7.1) can be rewritten as:

$$T = KI \tag{7.2}$$

i.e. the torque is proportional to the coil current and therefore the instrument scale is linear.

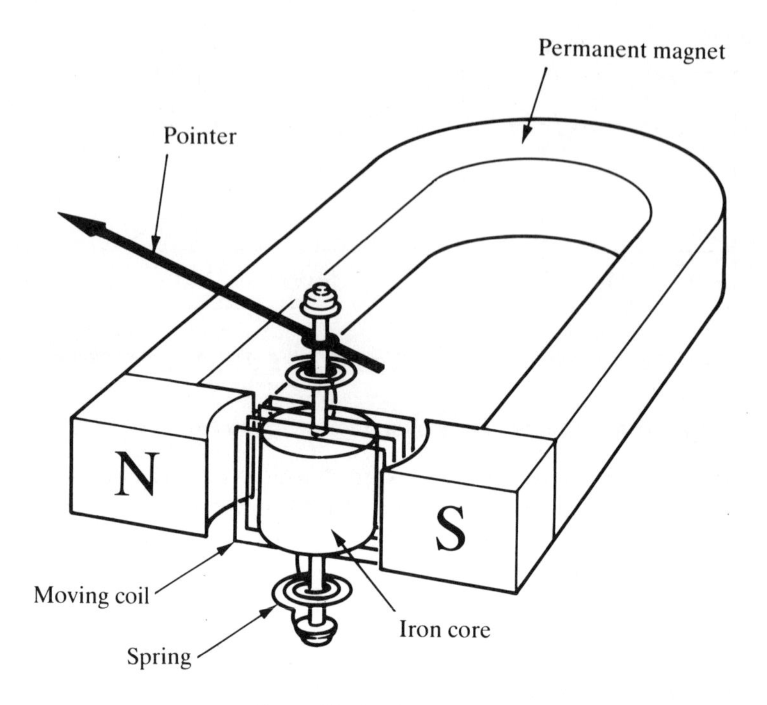

Figure 7.2 *Moving-coil meter*

In certain situations, a non-linear scale such as a logarithmic one is required, which is achieved by using either a specially shaped core or specially shaped magnet pole-faces.

While Figure 7.2 shows the traditional moving-coil instrument with a long U-shaped permanent magnet, many modern instruments employ much shorter magnets made from recently developed magnetic materials such as Alnico and Alcomax. These materials produce a substantially greater flux density, which, besides allowing the magnet to be smaller, has additional advantages in allowing reductions to be made in the size of the coil and in increasing the usable range of deflection of the coil to about 120°.

EXAMPLE 7.1

Calculate the reading which would be observed on a moving-coil ammeter when it is measuring the current with the waveform shown in Figure 7.3.

SOLUTION

A moving-coil meter measures the *mean* current, which is calculated from:

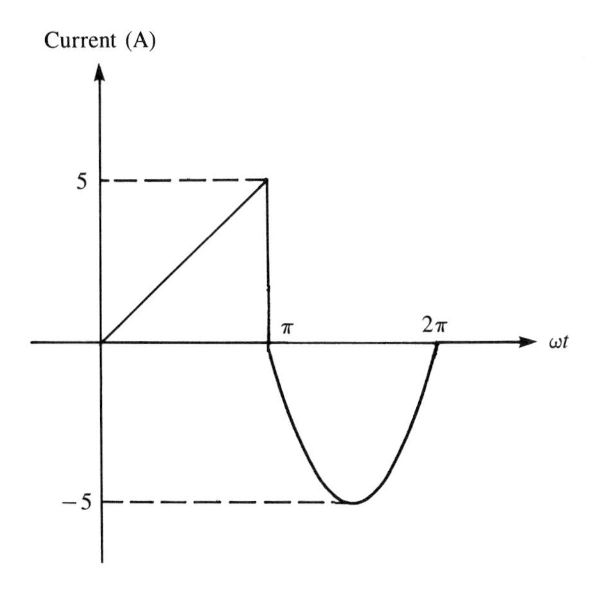

Figure 7.3 *Current waveform for examples 7.1 and 7.2*

$$
\begin{aligned}
I_{\text{mean}} &= \frac{1}{2\pi}\left[\int_0^{\pi}\frac{5\omega t}{\pi}\mathrm{d}\omega t + \int_{\pi}^{2\pi} 5\sin(\omega t)\mathrm{d}\omega t\right] \\
&= \frac{1}{2\pi}\left[\left[\frac{5(\omega t)^2}{2\pi}\right]_0^{\pi} + 5\left[-\cos(\omega t)\right]_{\pi}^{2\pi}\right] \\
&= \frac{1}{2\pi}\left[\frac{5\pi^2}{2\pi} - 0 - 5 - 5\right] \\
&= \frac{1}{2\pi}\left[\frac{5\pi}{2} - 10\right] = \frac{5}{2\pi}\left[\frac{\pi}{2} - 2\right] \\
&= -0.342\ \text{A}
\end{aligned}
$$

7.2.6 **Moving-iron meters**

This type of analog meter is suitable for measuring both d.c. and a.c. signals up to frequencies of 125 Hz. They are the cheapest form of meter available and consequently this type of meter is the one most commonly found in voltage and current measurement situations. The signal to be measured is applied to a stationary coil, and the associated field produced is often amplified by the presence of an iron structure associated with the fixed coil. The moving element in the instrument consists of an iron vane which is suspended within the field of the fixed coil. When the fixed coil is excited, a torque is generated and the iron vane turns in a direction which increases the

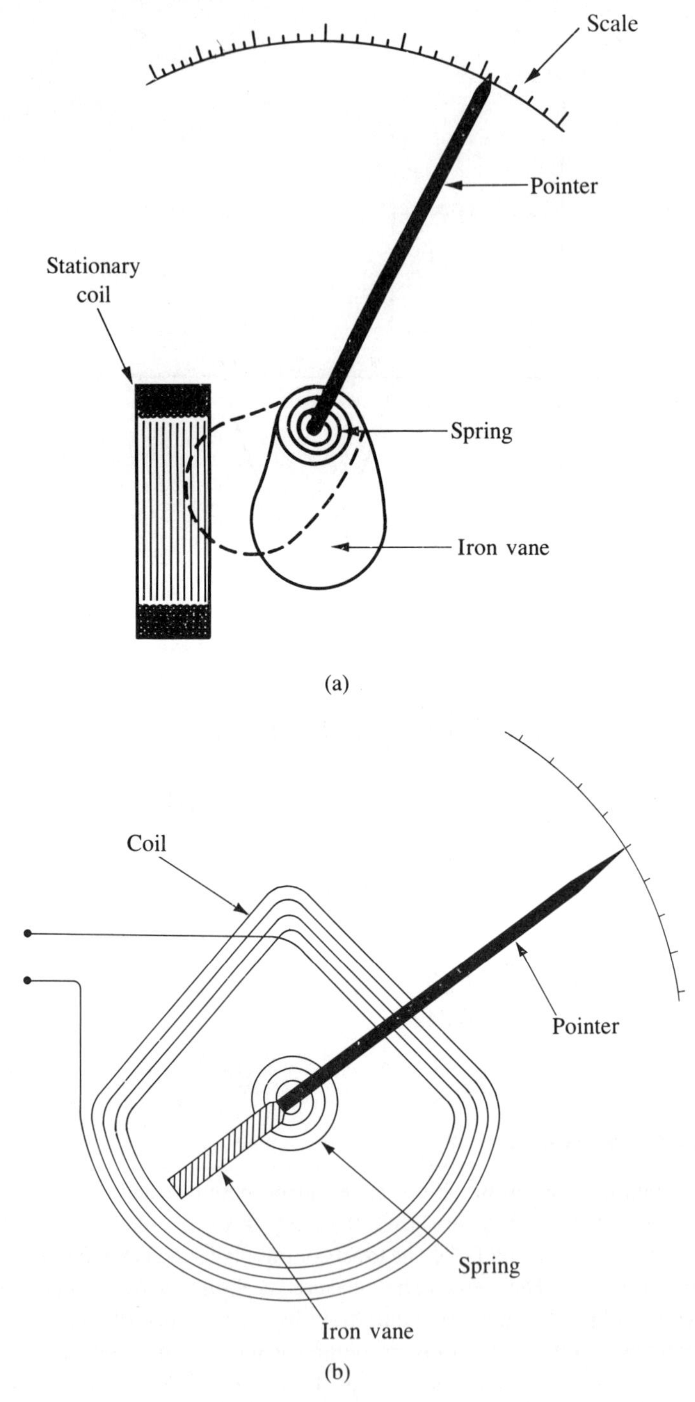

Figure 7.4 *(a) Attraction-type and (b) repulsion-type moving-iron meter*

flux in it. The majority of moving-iron instruments are either of the attraction or the repulsion type. A few instruments belong to a third combination type. The attraction type, where the iron vane is drawn into the field of the coil as the current is increased, is shown schematically in Figure 7.4(a). The repulsion type is sketched in Figure 7.4(b).

Provided that the iron components are operating below their magnetic saturation limits, the generated torque T produced in a moving iron meter in response to an excitation current I is given by:

$$T = \frac{I^2 \mathrm{d}M}{2\mathrm{d}\theta}$$

where M is the mutual inductance of the coil and vane and θ is the angular deflection of the pointer.

Rotation is opposed by a spring which produces a backwards torque T_s given by:

$$T_s = K\theta$$

where K is a constant. At equilibrium, $T = T_s$, and θ is therefore given by:

$$\theta = \frac{I^2 \mathrm{d}M}{2K\,\mathrm{d}\theta}$$

The instrument thus has a square-law response where the deflection is proportional to the square of the signal being measured, i.e. the output reading is a root-mean-squared (r.m.s.) quantity, and the output scale graduations are non-linear.

EXAMPLE 7.2

Calculate the reading which would be observed on a moving-iron ammeter when it is measuring the current whose waveform is shown in Figure 7.3.

SOLUTION

A moving iron meter measures the r.m.s. current I_{rms}, which is calculated from:

$$\begin{aligned}
{I_{rms}}^2 &= \frac{1}{2\pi}\left[\int_0^{\pi} \frac{25(\omega t)^2}{\pi^2}\mathrm{d}\omega t + \int_{\pi}^{2\pi} 25\sin^2(\omega t)\mathrm{d}\omega t\right] \\
&= \frac{1}{2\pi}\left[\int_0^{\pi} \frac{25(\omega t)^2}{\pi^2}\mathrm{d}\omega t + \int_{\pi}^{2\pi} \frac{25(1-\cos 2\omega t)}{2}\mathrm{d}\omega t\right] \\
&= \frac{25}{2\pi}\left[\left[\frac{(\omega t)^3}{3\pi^2}\right]_0^{\pi} + \left[\frac{\omega t}{2} - \frac{\sin 2\omega t}{4}\right]_{\pi}^{2\pi}\right] \\
&= \frac{25}{2\pi}\left[\frac{\pi}{3} + \frac{2\pi}{2} - \frac{\pi}{2}\right] \\
&= \frac{25}{2\pi}\left[\frac{\pi}{3} + \frac{\pi}{2}\right] = \frac{25}{2}\left[\frac{1}{3} + \frac{1}{2}\right] = 10.416\ \mathrm{A}^2
\end{aligned}$$

Therefore: $I_{rms} = ({I_{rms}}^2)^{1/2} = 3.23$ A.

7.2.7 Electrodynamic meters

Electrodynamic meters (or dynamometers) are analog instruments which can measure both d.c. and a.c. quantities up to a frequency of 2 kHz. The typical construction of an electrodynamic meter is illustrated in Figure 7.5. It consists of a circular moving coil mounted in the magnetic field produced by two separately wound circular stator coils which are connected in series with each other. The torque T is dependent upon the mutual inductance between the coils and is given by:

$$T = I_1 \cdot I_2 \cdot \frac{\mathrm{d}M}{\mathrm{d}\theta} \tag{7.3}$$

where I_1 and I_2 are the currents in the fixed and moving coils, respectively, M is the mutual inductance and θ represents the angular displacement between the coils.

When the instrument is used as an ammeter, the measured current is applied to both the moving coil and the fixed stator coils. The torque is thus proportional to the square of the current. If the measured current is alternating, the moving coil is unable to follow the alternating torque and it measures instead the mean value of the square of the current. By suitable graduation of the scale, the position of the pointer will show the square-root of this value, i.e. the r.m.s. current.

Electrodynamic meters are quite expensive but have the advantage of being quite accurate. Voltage, current and power can all be measured if the fixed and moving coils are connected appropriately.

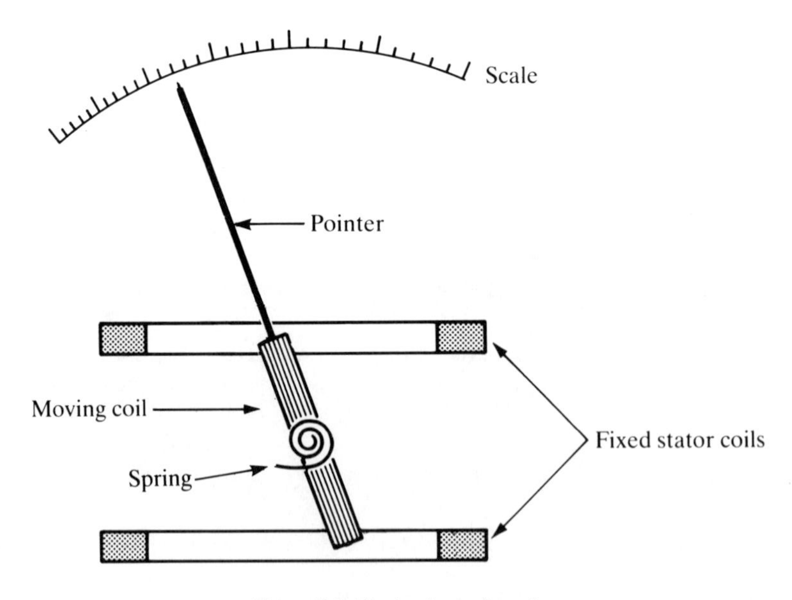

Figure 7.5 *Electrodynamic meter*

EXAMPLE 7.3

A dynamometer ammeter is connected in series with a 500 Ω resistor, a rectifying device and a 240 V r.m.s. alternating sinusoidal power supply. The rectifier behaves as a resistance of 200 Ω to current in one direction and as a resistance of 2 kΩ to current in the opposite direction. Calculate the reading indicated on the meter.

SOLUTION

$$V_{\text{peak}} = V_{\text{rms}}(2)^{1/2} = 339.4\,\text{V}$$

For $0 < \omega t < \pi$, $R = 700\,\Omega$ and for $\pi < \omega t < 2\pi$, $R = 2500\,\Omega$

$$\begin{aligned}
{I_{\text{rms}}}^2 &= \frac{1}{2\pi}\left[\int_0^{\pi}\frac{(339.4\sin\omega t)^2}{700^2}\,d\omega t + \int_{\pi}^{2\pi}\frac{(339.4\sin\omega t)^2}{2500^2}\,d\omega t\right] \\
&= \frac{339.4^2}{2\pi(10^4)}\left[\int_0^{\pi}\frac{\sin^2\omega t}{49}\,d\omega t + \int_{\pi}^{2\pi}\frac{\sin^2\omega t}{625}\,d\omega t\right] \\
&= \frac{339.4^2}{4\pi(10^4)}\left[\int_0^{\pi}\frac{(1-\cos 2\omega t)}{49}\,d\omega t + \int_{\pi}^{2\pi}\frac{(1-\cos 2\omega t)}{625}\,d\omega t\right] \\
&= \frac{339.4^2}{4\pi(10^4)}\left[\left(\frac{\omega t}{49} - \frac{\sin 2\omega t}{98}\right)_0^{\pi} + \left(\frac{\omega t}{625} - \frac{\sin 2\omega t}{1250}\right)_{\pi}^{2\pi}\right] \\
{I_{\text{rms}}}^2 &= \frac{339.4^2}{4\pi(10^4)}\left[\frac{\pi}{49} + \frac{\pi}{625}\right] = 0.0634\,\text{A}^2 \\
I_{\text{rms}} &= (0.0634)^{1/2} = 0.25\,\text{A}
\end{aligned}$$

7.2.8 *Induction meters*

Induction meters are analog instruments which respond only to alternating current signals. The meter consists of two electromagnets arranged such that a phase difference θ exists between the fluxes ϕ_1 and ϕ_2 associated with the two coils, as sketched in Figure 7.6. These two fluxes produce an induced e.m.f. and circulating concentric currents in the moving element of the instrument which consists of a conducting disk. The conducting disk tries to follow the rotating flux under the influence of a torque T given by:

$$T = K\phi_1\phi_2\sin\theta$$

For current and voltage measurement, the signal to be measured is applied to both electromagnetic coils, which results in the meter having a square-law output relationship of the form $T \propto I^2$ or $T \propto V^2$. The phase difference is obtained by shunting one coil with a non-inductive resistance, and rotation of the coil is opposed by a restraining spring.

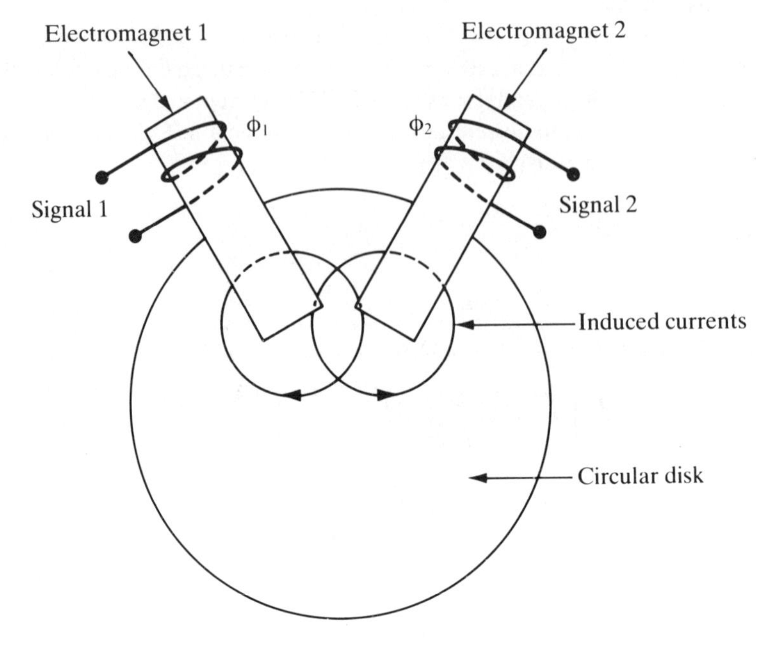

Figure 7.6 *Induction meter*

When used as an induction-type power meter, the disk in the instrument is unrestrained and free to rotate continuously. Circuit current is applied to one coil and circuit voltage to the other, thereby making the output reading a power measurement.

7.2.9 **Clamp-on meters**

These are used for measuring circuit currents and voltages in a non-invasive manner which avoids having to break the circuit being measured. The meter clamps onto a current-carrying conductor and the output reading is obtained by transformer action. The principle of operation is illustrated in Figure 7.7, where it can be seen that the clamp-on jaws of the instrument act as a transformer core and the current-carrying conductor acts as a primary winding. Current induced in the secondary winding is rectified and applied to a moving-coil meter. Although it is a very convenient instrument to use, the clamp-on meter has low sensitivity and the minimum current measurable is usually about one ampere.

7.2.10 **Electrostatic meters**

Electrostatic meters use the principle of attraction between two oppositely charged circular plates, as illustrated in Figure 7.8. The force produced is proportional to the square of the potential difference between the plates and is insufficient to provide an

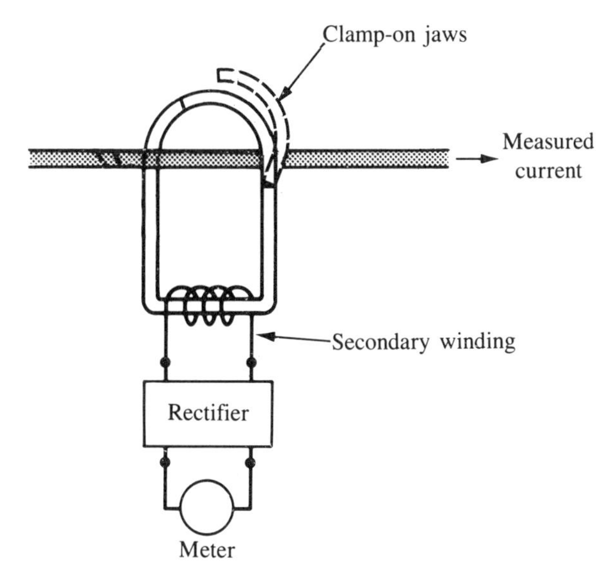

Figure 7.7 *Clamp-on meter*

adequate deflection at voltages less than about 50 V. Electrostatic meters are therefore normally only used for high-voltage applications, where voltages up to 500 kV can be measured directly. This type of instrument is very accurate, and it has the particular advantage of a very high input impedance, thereby drawing an extremely low current from the measured system.

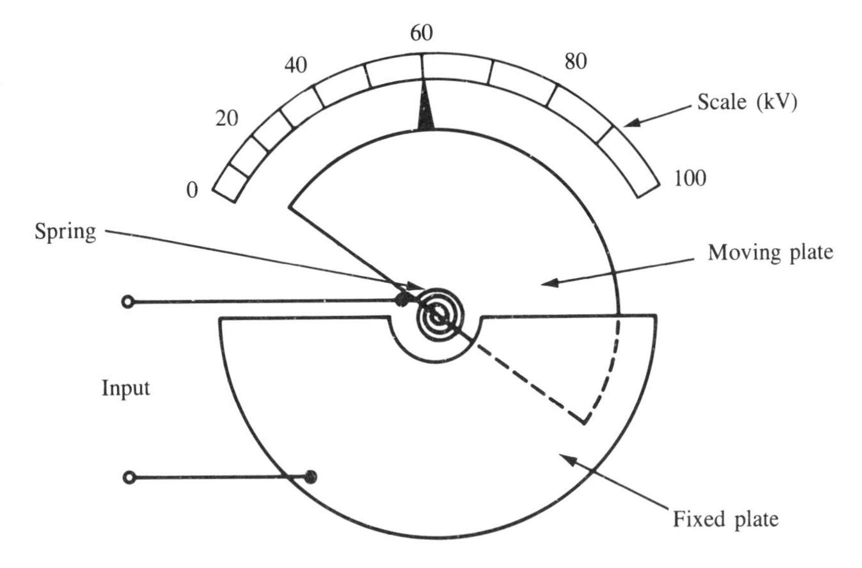

Figure 7.8 *Electrostatic meter*

7.3 Cathode-ray oscilloscope

The cathode-ray oscilloscope is probably the most versatile and useful instrument available for signal measurement. The more expensive models can measure signals at frequencies up to 500 MHz and even the cheapest models can measure signals up to 20 MHz. One particularly strong merit of the oscilloscope is its high input impedance, typically 1 MΩ, which means that the instrument has a negligible loading effect in most measurement situations. Some disadvantages of oscilloscopes include their fragility (being built around a cathode-ray tube) and their moderately high cost. Also, while the inaccuracy of the best instruments is only about ±1%, the inaccuracy of the cheapest ones can approach ±10%. The most important aspects in the specification of an oscilloscope are its bandwidth, its rise time and its accuracy.

The bandwidth is defined as the range of frequencies over which the oscilloscope amplifier gain is within 3 dB[1] of its peak value, as illustrated in Figure 7.9. The −3 dB point is where the gain is 0.707 times its maximum value. In most oscilloscopes, the amplifier is direct-coupled, which means that it amplifies d.c. voltages by the same factor as low-frequency a.c. ones. For such instruments, the minimum frequency measurable is zero and the bandwidth can be interpreted as the maximum frequency where the sensitivity (deflection/volt) is within 3 dB of the peak value. The oscilloscope must be chosen such that the maximum frequency to be measured is well within the bandwidth. The −3 dB specification means that an oscilloscope with a specified inaccuracy of ±2% and bandwidth of 100 MHz will have an inaccuracy of ±5% when measuring 30 MHz signals and this inaccuracy will increase still further at higher frequencies. Thus when applied to signal amplitude measurement, the oscilloscope is only usable at frequencies up to about 0.3 times its specified bandwidth.

The rise time is the transit time between the 10% and 90% levels of the response when a step input is applied to the oscilloscope. Oscilloscopes are normally designed such that:

$$\text{bandwidth} \times \text{rise time} = 0.35$$

Thus, for a bandwidth of 100 MHz, rise time $= 0.35/100 \times 10^6 = 3.5$ ns.

An oscilloscope is a relatively complicated instrument which is constructed from a number of subsystems, and it is necessary to consider each of these in turn in order to understand how the complete instrument functions.

7.3.1 Cathode-ray tube

The cathode-ray tube is the fundamental part of an oscilloscope and is shown in Figure 7.10. The cathode consists of a barium and strontium oxide coated, thin, heated

[1] The decibel, commonly written dB, is used to express the ratio between two quantities. For two voltage levels V_1 and V_2, the difference between the two levels is expressed in decibels as: $20\log_{10}(V_1/V_2)$. It follows from this that: $20\log_{10}(0.707/1) = -3$ dB.

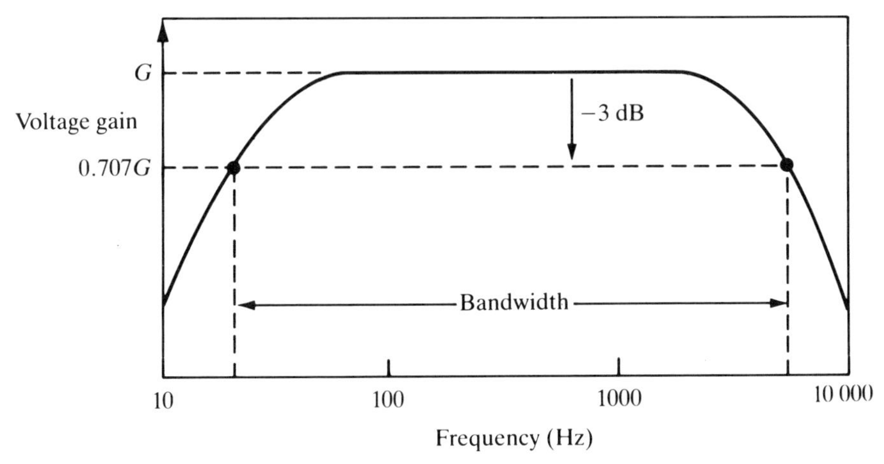

Figure 7.9 *Bandwidth*

filament from which a stream of electrons is emitted. The stream of electrons is focused onto a well-defined spot on a fluorescent screen by an electrostatic focusing system which consists of a series of metal disks and cylinders charged at various potentials. Adjustment of this focusing mechanism is provided by controls on the front panel of the oscilloscope. An 'INTENSITY' control varies the cathode heater current and therefore the rate of emission of electrons, and thus adjusts the intensity of the display on the screen. These and other typical controls are shown in the illustration of the front panel of a simple oscilloscope given in Figure 7.11.

Application of electrostatic potentials to two sets of deflector plates mounted at right angles to one another within the tube provide for deflection of the stream of electrons, such that the spot where the electrons are focused on the screen is moved. The two sets of deflector plates are normally known as the horizontal and vertical

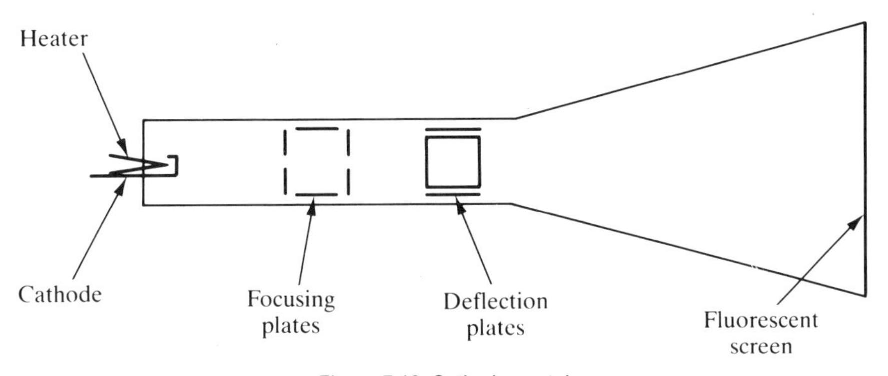

Figure 7.10 *Cathode-ray tube*

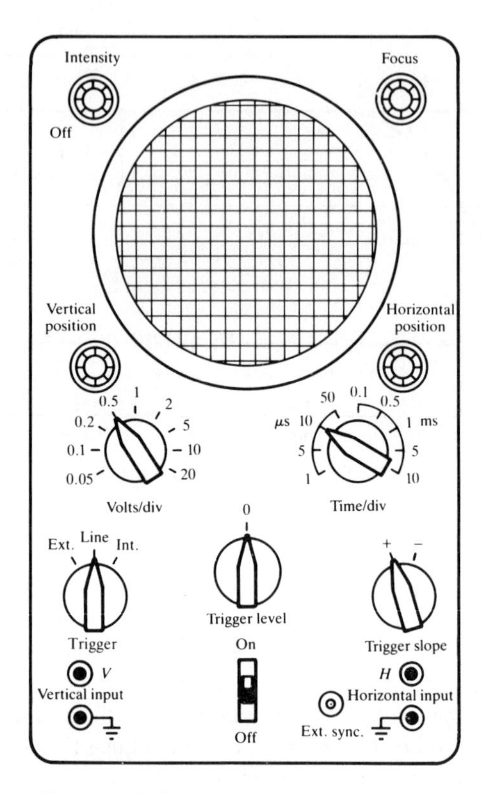

Figure 7.11 ***Controls of a simple oscilloscope***

deflection plates, according to the respective motion caused to the spot on the screen. The magnitude of any signal applied to the deflector plates can be calculated by measuring the deflection of the spot against a graticule etched on a glass plate in front of the tube screen.

In the oscilloscope's most common mode of use measuring time-varying signals, the unknown signal is applied, via an amplifier, to the y-axis (vertical) deflector plates and a time-base signal to the x-axis (horizontal) deflector plates. In this mode of operation, the display on the oscilloscope screen is in the form of a graph with the magnitude of the unknown signal on the vertical axis and time on the horizontal axis.

The intensity of the display on the screen of an oscilloscope decays with time. However, a repeating signal can be refreshed periodically to produce a stationary, non-fading pattern on the screen by triggering the signal at an appropriate rate with a *time-base circuit*, as described later in this section.

If signals are non-repeating, for example a signal derived from the sound wave following an explosion, special storage oscilloscopes must be used to display and analyze the signal. *Analog storage oscilloscopes* use a special phosphorescent coating on the screen which can 'hold' a trace for up to one hour. Alternatively, *digital storage oscilloscopes* are now widely used for this purpose. These consist of a conventional analog oscilloscope which has digital storage capabilities. The analog signal is converted and stored in digital format and then reconverted to analog form at the necessary frequency to refresh the analog display on the screen and so obtain a non-fading display of the signal.

7.3.2 **Channels**

The basic subsystem, of an electron source, focusing system and deflector plates, is often duplicated one or more times within the cathode-ray tube to provide a capability of displaying two or more signals together on the screen. Each such basic subsystem is known as a channel, and therefore the commonest oscilloscope configuration having two channels can display two separate signals simultaneously.

7.3.3 **Single-ended inputs**

Most simple oscilloscopes have this type of input, which only allows signal voltages to be measured relative to ground. This type can be recognized by the presence of only one input terminal plus a ground terminal per oscilloscope channel.

7.3.4 **Differential inputs**

This type of input is provided on more expensive oscilloscopes and allows the potentials at two non-grounded points in a circuit to be compared. Two input terminals plus a ground terminal are provided for each channel. This arrangement can also be used in single-ended mode to measure a signal relative to ground by using just one of the input terminals plus ground.

7.3.5 **Input coupling switch**

This is available on many oscilloscopes and alters the manner in which the input signal is connected to the input channel. Normally, three positions are available, d.c., a.c. and ground. With the switch in the 'a.c.' position, a capacitor or other high-pass filter is connected across the input. This is used to remove the d.c. component in signals where the a.c. component is of small magnitude compared with the d.c. level. Some distortion of the signal may result from this filtering. This is avoided in the 'd.c.' position, which does not condition the input signal at all and passes all frequency components. The 'ground' switch position connects the input signal to the oscilloscope chassis, providing a convenient way of establishing the zero voltage level.

7.3.6 **Time-base circuit**

The purpose of a time-base is to apply a voltage to the horizontal deflector plates such that the horizontal displacement of the spot is proportional to time. This voltage, in the form of a ramp known as a sweep waveform, must be applied repetitively, such that the motion of the spot across the screen appears as a straight line when a d.c. level is applied to the input channel. Furthermore, this time-base voltage must be synchronized with the input signal in the general case of a time-varying signal, such that a steady trace is obtained on the oscilloscope screen. The length of time taken for the spot to traverse the screen is controlled by a 'TIME/DIV' switch, which sets the length of time taken by the spot to travel between two marked divisions on the screen, thereby allowing signals at a wide range of frequencies to be measured.

Each cycle of the sweep waveform is initiated by a pulse from a pulse generator. The pulse generator is controlled by a triggering signal, with a pulse being generated every time the triggering signal crosses a preselected slope and voltage level condition. This condition is defined by the 'TRIGGER LEVEL' and 'TRIGGER SLOPE' switches. The former selects the voltage level on the trigger signal, commonly zero, at which a pulse is generated, while the latter selects whether pulsing occurs on a positive or negative going part of the triggering waveform.

Synchronization of the sweep waveform with the measured signal is most easily achieved by deriving the trigger signal from the measured signal, a procedure which is known as *internal triggering*. Alternatively, *external triggering* can be applied if the frequencies of the triggering signal and measured signals are related by an integer constant such that the display is stationary. External triggering is necessary when the amplitude of the measured signal is too small to drive the pulse generator, and it is also used in applications where there is a requirement to measure the phase difference between two sinusoidal signals of the same frequency. It is very convenient to use the 50 Hz line voltage for external triggering when measuring signals at mains frequency, and this is often given the name *line triggering*.

7.3.7 **Vertical sensitivity control**

In order to make the range of signal magnitudes measurable by an oscilloscope as wide as possible, a series of attenuators and pre-amplifiers are provided at the input. These condition the measured signal to the optimum magnitude for input to the main amplifier and vertical deflection plates. Selection of the appropriate input amplifier/attenuator is made by setting a 'VOLTS/DIV' control associated with each oscilloscope channel. This defines the magnitude of the input signal which will cause a deflection of one division on the screen.

7.3.8 **Display position control**

The position on the screen at which a signal is displayed can be controlled in two ways. The horizontal position can be adjusted by a 'HORIZONTAL POSITION' knob on the oscilloscope front panel and similarly a 'VERTICAL POSITION' knob controls the vertical position. These controls adjust the position of the display by biasing the measured signal with d.c. voltage levels.

7.4 Voltage measurement

Atoms are composed of electrically charged particles, called electrons and protons, which occur in equal numbers in the equilibrium state. If a deficit in either type of particle exists, the atom is said to have either a positive or negative charge, which is measured in units called *coulombs* (C). Such a charge is a source of potential energy and it has the capacity to do work. If such a charge is present at some position within an electrical circuit, it will move to some lower energy level if released, and expend energy in so doing. This net positive or negative charge is called a *potential difference* or *voltage* and is measured in units called *volts* (V).

The particular concern in this chapter is not with means of generating voltages, but merely with ways of measuring them. The measuring instruments available include analog electromechanical voltmeters, analog electronic voltmeters, the cathode-ray oscilloscope and digital voltmeters. The first of these, electromechanical voltmeters, are a passive type of instrument, but all the others are active instruments which include signal amplification stages of some form.

The electromechanical group consists of the moving-iron voltmeter, the moving-coil voltmeter, the electrostatic voltmeter, the dynamometer voltmeter and the thermocouple meter. Most of these were discussed in detail in section 7.2 and therefore the coverage of them in the following paragraphs is brief.

Electronic instruments with an analog form of output have some important advantages. Firstly, they have a high input impedance which avoids the circuit-loading problems associated with many applications of electromechanical instruments. Secondly, they have an amplification capability which enables them to measure small signal levels accurately. D.C. electronic voltmeters exist as both direct-coupled and chopper-amplifier types. In the case of a.c. versions, three different types exist, known as average-responding, peak-responding and r.m.s.-responding, respectively.

Digital voltmeters have been developed to satisfy a need for higher measurement accuracies and a faster speed of response to voltage changes. As a voltage-indicating instrument, the binary nature of the output reading is readily applied to a display in the form of discrete numerals. Where human operators are required to monitor system voltage levels, this form of output makes an important contribution to measurement reliability and accuracy, since the problem of meter parallax error is eliminated and the possibility of gross error through misreading the meter output is greatly reduced. The availability of a direct output in digital form has also become very valuable to the rapidly expanding range of computer-control applications. Digital voltmeters differ mainly in the principle used to effect the analog-to-digital conversion between the measured analog voltage and the output digital reading.

A further class of instruments exists for measuring the very high voltages associated with the electricity supply grid system and certain other systems. The applications for these are so specialized that it is not appropriate to discuss them within this text and the reader is referred elsewhere if such information is required (Buckingham and Price, 1966).

7.4.1 **Electromechanical voltmeters**

The very earliest instruments used for measuring voltage relied on the principle of interaction between charged bodies. This idea was exploited by Kelvin in the nineteenth century in the development of the electrostatic voltmeter, which measures the attraction forces between two oppositely charged plates. Such forces are relatively small, and consequently this instrument is now only used for measuring d.c. voltages greater than 100 volts in magnitude.

The most popular d.c. analog voltmeter used today is the moving-coil type, because of its sensitivity, accuracy and linear scale. This operates at low current levels of one milliamp or so, and hence the basic instrument is only suitable for measuring voltages in the millivolt range and up to around two volts. For higher voltages, the measuring range of the instrument is increased by placing a resistance in series with the coil, such that a known proportion of the applied voltage is measured by the meter.

Alternative instruments for measuring d.c. voltages are the dynamometer voltmeter and the moving-iron meter. In their basic form, these instruments are most suitable for measuring voltages in the range of 0–30 volts. Higher voltages are measured by placing a resistance in series with the instrument, as in the case of the moving-coil type.

For measuring a.c. voltages, the usual choice is between the dynamometer voltmeter and the moving-iron meter. Both of these have an r.m.s. form of output. Of the two, the dynamometer voltmeter is more accurate but also more expensive. In both cases, range extension is obtained by using the instruments with a series resistance. This is beneficial to a.c. voltage measurements, as it compensates for the effect of coil inductance by reducing the total resistance/inductance ratio, and hence measurement accuracy is improved. For measurements above about 300 volts, it becomes impractical to include a suitable resistance within the case of the instrument because of heat-dissipation problems, and instead an external resistance is used.

For measurement of high a.c. voltages, an alternative to using one of these instruments with very large resistances is to transform the voltage to a lower value before applying the measuring instrument. This procedure is the recommended industrial practice for measuring voltages above 750 volts.

One major limitation in applying analog meters to a.c. voltage measurement is that the maximum frequency measurable directly is low, 2 kHz for the dynamometer voltmeter and only 100 Hz in the case of the moving iron instrument. A partial solution to this limitation is to rectify the voltage signal and then apply it to a moving-coil meter, as shown in Figure 7.12. This extends the upper measurable frequency limit to 20 kHz. However, the inclusion of the bridge rectifier makes the measurement system temperature-sensitive, and non-linearities significantly affect measurement accuracy for voltages which are small relative to the full-scale value.

An alternative solution to the upper frequency limitation is provided by the *thermocouple meter*, shown in Figure 7.13. In this, the a.c. voltage signal heats a small element, and the resulting temperature rise is measured by a thermocouple. The d.c. voltage generated in the thermocouple is applied to a moving-coil meter. The output

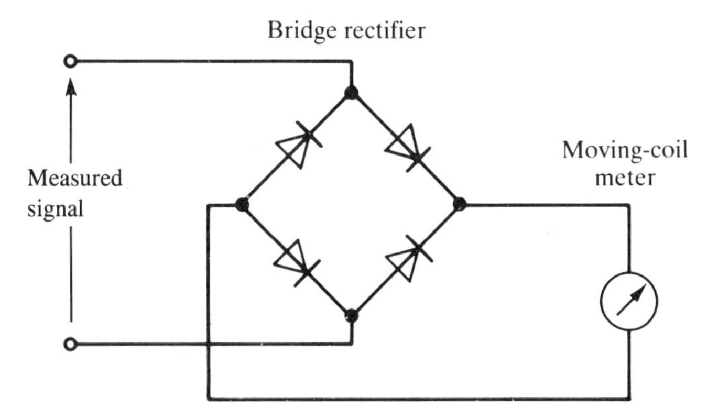

Figure 7.12 *Measurement of high-frequency voltage signals*

meter reading is an r.m.s. quantity which varies in a non-linear fashion with the magnitude of the measured voltage. Very high frequencies up to 50 MHz can be measured by this method.

Before ending this section on electromechanical meters, mention must also be made of the *analog multimeter*, whose working arrangement is shown in Figure 7.14. This combines a d.c. moving-coil meter with a bridge rectifier to extend its measuring capability to a.c. quantities. A set of rotary switches allows the selection of various series and shunt resistors, which make the instrument capable of measuring both voltage and current over a number of ranges. An internal power source is also often provided to allow it to measure resistances as well. Whilst this instrument is very useful for giving an indication of voltage levels, the compromises in its design which enable it to measure so many different quantities necessarily mean that its accuracy is not as good as instruments which are purpose-designed to measure just one quantity over a single measuring range.

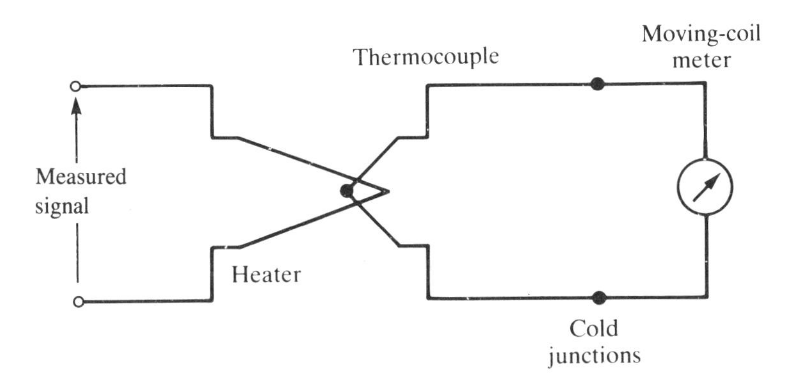

Figure 7.13 *The thermocouple meter*

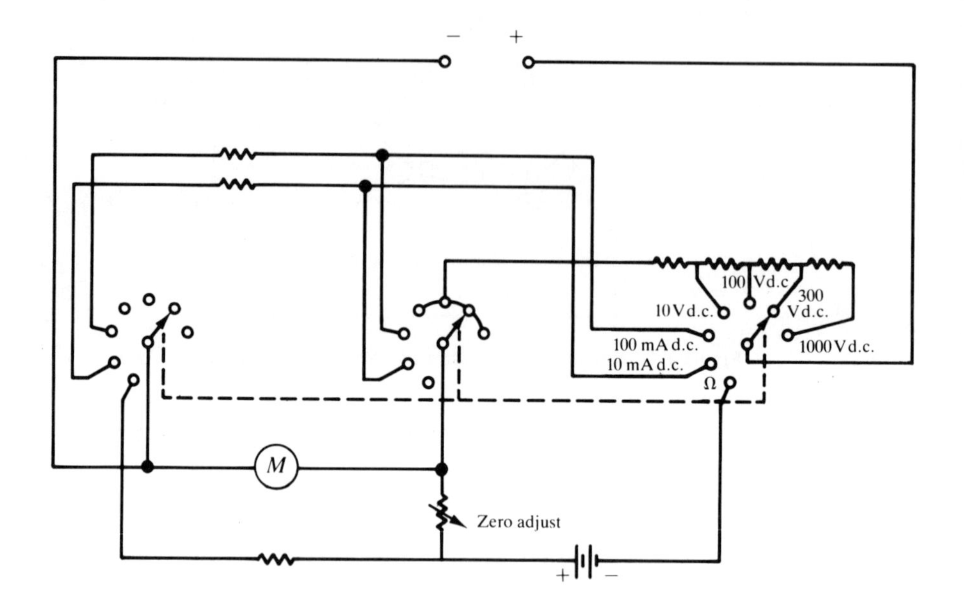

Figure 7.14 *The analog multimeter*

7.4.2 **Electronic analog voltmeters**

The standard electronic voltmeter for d.c. measurements consists of a simple direct-coupled amplifier and a moving-coil meter, as shown in Figure 7.15(a). For measurement of very low-level voltages of a few microvolts, a more sophisticated circuit, known as a chopper amplifier, is used. This is shown in Figure 7.15(b). In this, the d.c. input is chopped at a low frequency of around 250 Hz, passed through a blocking capacitor, amplified, passed through another blocking capacitor to remove drift, demodulated, filtered and applied to a moving-coil meter.

Three versions of electronic voltmeter exist for measuring a.c. signals. The *average-responding type* is essentially a direct coupled d.c. electronic voltmeter with an additional rectifying stage at the input. The output is a measure of the average value of the measured voltage waveform. The second form, known as a *peak-responding type*, has a half-wave rectifier at the input followed by a capacitor. The final part of the circuit consists of an amplifier and moving-coil meter. The capacitor is charged to the peak value of the input signal, and therefore the amplified signal applied to the moving-coil meter gives a reading of the peak voltage in the input waveform. Finally, a

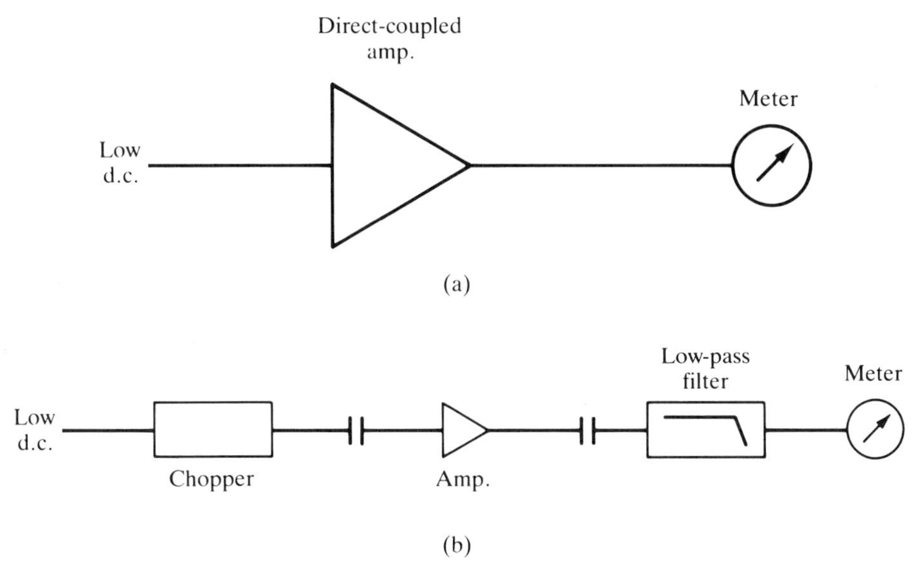

Figure 7.15 *(a) Simple d.c. electronic voltmeter; (b) d.c. electronic voltmeter with chopper amplifier*

third type is available, known as an *r.m.s.-responding type*, which gives an output reading in terms of the r.m.s. value of the input waveform. This type is essentially a thermocouple meter in which an amplification stage has been inserted at the input.

7.4.3 Cathode-ray oscilloscope

The principles of operation of the cathode-ray oscilloscope have already been discussed in section 7.3. The instrument is widely used for voltage measurement, especially as an item of test equipment for circuit fault-finding. It is not a particularly accurate instrument, and is best used where only an approximate measurement is required. Its great value lies in its versatility and applicability to measuring a very wide range of a.c. and d.c. voltages. As a test instrument, it is often required to measure voltages whose frequency and magnitude are totally unknown. The set of rotary switches which alter its range and time-base so easily, and the circuitry which protect it from damage when high voltages are applied to it on the wrong range, make it ideally suitable for such applications.

7.4.4 Digital voltmeter (DVM)

The major part of the digital voltmeter is the circuitry that converts the analog voltage being measured into a digital quantity. Various techniques are used to effect the analog-to-digital conversion. As a general rule, the more expensive and complicated conversion methods achieve a faster conversion speed.

As the instrument only measures d.c. quantities in its basic mode, another necessary component within it is one that performs a.c.–d.c. conversion and thereby gives it the capacity to measure a.c. signals. After conversion, the voltage value is displayed by means of indicating tubes or a set of solid-state light-emitting diodes. Four-, five- or even six-figure output displays are commonly used, and although the instrument itself may not be inherently more accurate than some analog types, this form of display enables measurements to be recorded with much greater resolution than that obtainable by reading an analog meter scale.

The simplest form of DVM is the ramp type. Its main drawbacks are non-linearities in the shape of the ramp waveform used and lack of noise rejection, and these problems mean typical inaccuracy levels are ±0.05%. It is relatively cheap, however. Another cheap form of DVM is the potentiometric type, and this gives excellent performance for its price. Better noise-rejection capabilities and correspondingly less measurement inaccuracy (down to ±0.005%) are obtained with integrating-type DVMs. Unfortunately, these are quite expensive.

The digital multi-meter is an extension of the DVM which can measure a.c. and d.c. voltages over a number of ranges through inclusion within it of a set of switchable amplifiers and attenuators. It is widely used in circuit test applications as an alternative to the analog multi-meter, and includes protection circuits which prevent damage if high voltages are applied to the wrong range.

7.5 **Current measurement**

If atoms with a net positive or negative charge are released from some point in an electrical circuit, the electrically charged particles will flow away from the point of unequal charge distribution until an equilibrium state is reached. The source of charged particles is known as a potential difference, and their rate of flow is known as a current. If charged particles flow in opposite directions along two parallel conductors, a force is exerted, and this phenomenon is used to define the unit of *amperes* (A) by which current is measured. When two currents, of magnitude one ampere each, flow in opposite directions along two infinitely long parallel wires positioned one metre apart in vacuum, the force produced on each conductor is 2×10^{-7} newtons per metre length of conductor.

Any of the current-operated electromechanical voltmeters discussed in section 7.2 can be used to measure current by placing them in series with the current-carrying circuit, and the same frequency limits apply to the measured signal. As in the case of voltage measurement, this upper frequency limit can be raised by rectifying the a.c. current prior to measurement or by using a thermocouple meter. To minimize the effect on the measured system, any current-measuring instrument should have a small resistance. This is the opposite of the case of voltage measurement where the instrument is required to have a high resistance for minimal circuit loading.

Moving-coil meters are suitable for direct measurement of d.c. currents in the milliamp range up to one ampere and the range for dynamometer ammeters is up to a few amperes. Moving iron meters can measure d.c. currents up to several hundred amperes directly. Similar measurement ranges apply when moving iron and dynamometer-type instruments are used to measure a.c. currents.

To measure currents higher than allowed by the basic design of electromechanical instruments, it is necessary to insert a shunt resistance into the circuit and measure the potential difference across it. One difficulty which results from this technique is the large value of power in the shunt when high-magnitude currents are measured. In the case of a.c. current measurement, care must also be taken to match the resistance and reactance of the shunt to that of the measuring instrument so that frequency and waveform distortion in the measured signal are avoided.

Current transformers provide an alternative method of measuring high-magnitude currents, which avoids the difficulty of designing a suitable shunt. Different versions of these exist for transforming both d.c. and a.c. currents. A d.c. current transformer is shown in Figure 7.16. The central d.c. conductor in the instrument is threaded through two magnetic cores which carry two high-impedance windings connected in series opposition. It can be shown (Baldwin, 1973) that the current in the windings when excited with an a.c. voltage is proportional to the d.c. current in the central conductor. This output current is commonly rectified and then measured by a moving-coil instrument.

An a.c. current transformer typically has a primary winding consisting of only a few copper turns wound on a rectangular or ring-shaped core. The secondary winding on the other hand would normally have several hundred turns according to the current step-down ratio required. The output of the secondary winding is measured by any suitable current-measuring instrument. The design of current transformers is substantially different from that of voltage transformers. The rigidity of its mechanical construction has to be sufficient to withstand the large forces arising from short-circuit currents, and special attention has to be paid to the insulation between its windings for similar reasons. A low-loss core material is used and flux densities are kept as small as possible to reduce losses. In the case of very high currents, the primary winding often

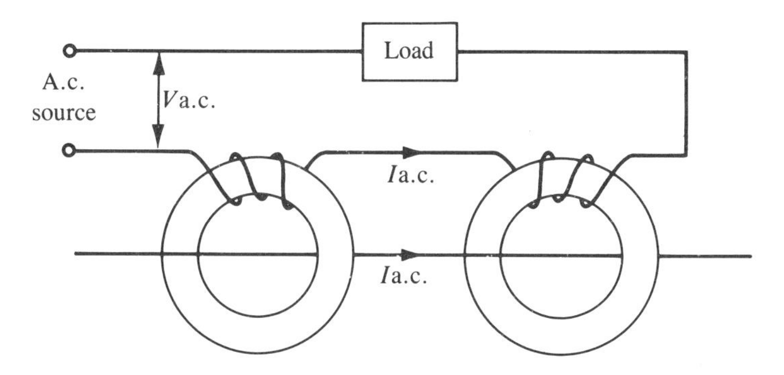

Figure 7.16 *D.C. current transformer*

consists of a single copper bar which behaves as a single-turn winding. The clamp-on meter, described in section 7.2, is a good example of this.

Apart from electromechanical meters, all the other instruments for measuring voltage discussed in section 7.2 can be applied to current measurement by using them to measure the voltage drop across a known resistance placed in series with the current-carrying circuit. The digital voltmeter and electronic meters are widely applied for measuring currents accurately by this method, and the cathode-ray oscilloscope is frequently used to obtain approximate measurements in circuit-test applications. Finally, mention must also be made of the use of digital and analog multi-meters for current measurement, particularly in circuit-test applications. These instruments include a set of switchable dropping resistors and so can measure currents over a wide range. Protective circuitry within such instruments prevents damage when high currents are applied on the wrong input range.

7.6 **Resistance measurement**

The standard methods available for measuring resistance, which is measured in units of ohms (Ω), include the voltmeter–ammeter method, the resistance-substitution method, the ohmmeter and the d.c. bridge circuit. The digital voltmeter can also be used for measuring resistance if an accurate current source is included within it which establishes a current in the resistance. This can give a measurement uncertainty of ±0.1%. Apart from the ohmmeter, these instruments are normally only used to measure medium-value resistances in the range of 1 Ω to 1 MΩ. Special instruments are available for obtaining high-accuracy resistance measurements outside this range (see Baldwin, 1973).

7.6.1 ***Voltmeter–ammeter method***

The voltmeter–ammeter method consists of applying a measured d.c. voltage across the unknown resistance and measuring the current in the resistance. Two alternatives exist for connecting the two meters, as shown in Figure 7.17. In Figure 7.17(a), the ammeter measures the current in both the voltmeter and the resistance. The error due to this is minimized when the measured resistance is small relative to the voltmeter resistance. In the alternative form of connection, Figure 7.17(b), the voltmeter measures the voltage drop across the unknown resistance and the ammeter. Here, the measurement error is minimized when the unknown resistance is large with respect to the ammeter resistance. Thus method (a) is best for measurement of small resistances and method (b) for large ones.

Having thus measured the voltage and current, the value of the resistance is then calculated very simply by applying Ohm's law. This is a suitable method wherever the measurement uncertainty of ±1% which it gives is acceptable.

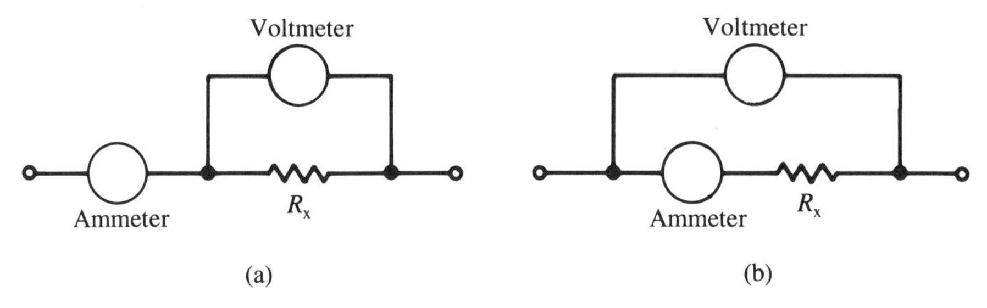

Figure 7.17 *Voltmeter–ammeter method of measuring resistance*

7.6.2 Resistance-substitution method

In the voltmeter–ammeter method above, either the voltmeter is measuring the voltage across the ammeter as well as across the resistance, or the ammeter is measuring the current in the voltmeter as well as in the resistance. The measurement error caused by this is avoided in the resistance-substitution technique. In this method, an unknown resistance in a circuit is temporarily replaced by a variable resistance of known value. The variable resistance is adjusted until the measured circuit voltage and current are the same as those that existed with the unknown resistance in place. The variable resistance at this point is equal in value to the unknown resistance.

7.6.3 Ohmmeter

The ohmmeter is a simple instrument in which a battery applies a known voltage V_b across a combination of the unknown resistance R_u and a known resistance R_1 connected in series, as shown in Figure 7.18. Measurement of the voltage V_m across the known resistance R_1 allows the unknown resistance, R_u, to be calculated from:

$$R_u = \frac{R_1(V_b - V_m)}{V_m}$$

Ohmmeters are used to measure resistances over a wide range from a few milliohms up to 50 MΩ. The lowest measurement inaccuracy is ±2% and ohmmeters are therefore more suitable for use as test equipment rather than in applications where high accuracy is required. Most of the available versions contain a switchable set of standard resistances, so that measurements of reasonable accuracy over a number of ranges can be made.

Most *digital and analog multi-meters* contain circuitry of the same form as in an ohmmeter, and hence can be similarly used to obtain approximate measurements of resistance.

7.6.4 D.C. bridge circuit

D.C. bridge circuits, as discussed in Chapter 6, provide the most commonly used

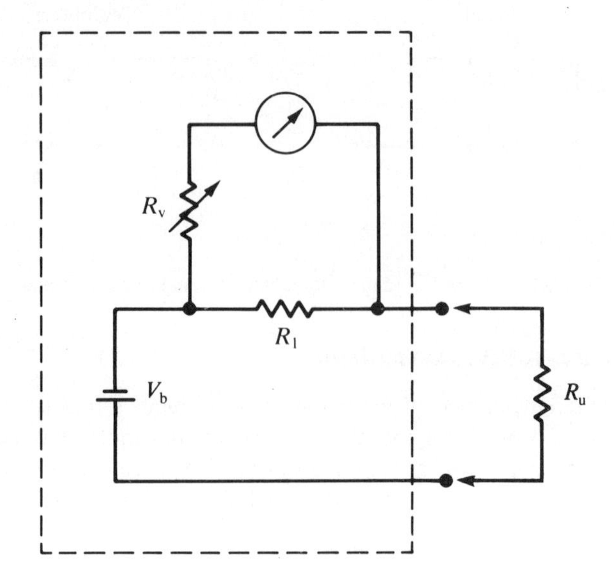

Figure 7.18 *The ohmmeter*

method of measuring medium value resistances. The lowest measurement inaccuracy is provided by the null-output type Wheatstone bridge and figures of ±0.02% are achievable with commercially available instruments. Deflection-type bridge circuits are simpler to use in practice than the null-output type, but their measurement accuracy is inferior and the non-linear output relationship is an additional difficulty. One advantage they have is in enabling resistance measurements to be used as inputs to automatic control schemes.

7.7 **Power measurement**

Measurement of the power in an electrical circuit is needed for many reasons. Circuit components have to be designed to have an adequate power-handling capability, and appropriate means of heat dissipation have to be provided. Such measures are necessary to protect equipment from damage and also to avoid fire hazards due to equipment overheating. Measurement of both the useful power and the power losses in a circuit also allows power utilization efficiency to be optimized.

Power measurement is also important to the electricity supply authorities, to ensure that the energy supplied to customers is properly costed and charged for. It is also required to ensure that the total system load is within the capacity of the available generating plant.

Power is measured in units of *watts* (W). The energy required to move one coulomb of charge between two energy levels with a potential difference equal to one volt is called a *joule* (J), and the rate at which this is done is a measure of power expressed in joules per second or watts.

In the case of d.c. circuits, the power is the simple product of the measured voltage and current. In the case of a.c. circuits, where the voltage and current are varying cyclically but often with a phase difference, it is possible to define a quantity called the instantaneous power which is the product of the voltage level and current level at any instant of time. The instantaneous power varies in a cyclic manner, however, and is of little practical value. Power in a.c. circuits is therefore normally quantified by the average power in watts, which is calculated according to:

$$P_{avg} = V_{rms} I_{rms} \cos(\phi)$$

where ϕ is the phase difference between the current and voltage waveforms.

It is appropriate in some circumstances to measure the reactive power as well, in units of 'volt amperes reactive' (VA_r). This is given by:

$$P_{reactive} = V_{rms} I_{rms} \sin(\phi)$$

These two quantities are related to the instantaneous power by the vector diagram shown in Figure 7.19.

The two common instruments used for measuring power are the dynamometer wattmeter and the electronic wattmeter. Certain other instruments are also used for measurement of high power levels in special applications, such as within the electricity supply industry, as described in Wolf (1973).

7.7.1 Dynamometer wattmeter

The dynamometer wattmeter is the standard instrument used for most power measurements at frequencies below 400 Hz. Products of voltages up to 300 volts and currents up to 20 amperes can be measured. The instrument is the same standard type

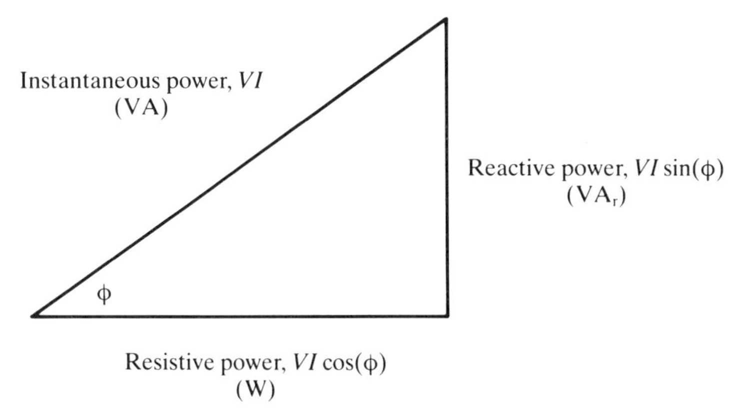

Figure 7.19 *Vector diagram showing relationship between resistive and reactive power components*

of dynamometer described in section 7.2 for measuring voltage, except that the coils are connected differently. When used for measuring power dissipation, the stationary coil is connected in series with the measured circuit and carries the circuit current. The rotating coil has a large resistance placed in series with it and is connected across the measured circuit, thus measuring the voltage drop. Theoretically, the instrument measures the product of instantaneous voltage and current and therefore the output pointer reading could be expected to be a measure of instantaneous power, which would make it oscillate at a high frequency. In practice, system inertia prevents such oscillation if the measured circuit has a frequency greater than a few hertz and the pointer settles to a reading proportional to the average circuit power.

Measurement inaccuracy in commercial instruments varies between ±0.1% and ±0.5% of full scale, according to the design and cost of the instrument. A major source of error in dynamometer wattmeters is power dissipation in the coils. The error due to this becomes very large when the power level in the measured circuit is low. Consequently, the dynamometer wattmeter is not suitable for measuring power levels below 5 watts.

Apart from measuring power dissipation, the dynamometer can also measure both the reactive power and the power factor, cos (ϕ), in a circuit. Modifications to the coils and their mode of connection is necessary for this as described in Wolf (1973).

7.7.2 **The electronic wattmeter**

The electronic wattmeter is capable of measuring power levels between 0.1 W and 100 kW in circuits operating at frequencies up to 100 kHz. Below 40 kHz, its inaccuracy is ±3%, but at higher frequencies inaccuracy can be as high as ±10%. The instrument is used wherever its accuracy is superior to that of the dynamometer, i.e. at power levels below 5 watts and at all frequencies above 400 Hz.

The basic operation of such instruments relies on a circuit containing two amplifiers. One amplifier measures the current in the load and the other measures the voltage across the load. The amplifier outputs are multiplied together by a circuit which uses the logarithmic input relationship of a transistor. The multiplied output is proportional to the instantaneous power level, and is converted to a reading of the average power by the inertial properties of the moving-coil meter to which it is applied. The moving-coil meter ceases to average the instantaneous power properly below a certain frequency, and consequently the electronic wattmeter is not used below frequencies of 10 Hz.

7.7.3 **Power measurement in three-phase circuits**

For power measurement in three-phase circuits, two wattmeters are required, as shown in Figure 7.20. The total average power is then given by the readings on the two wattmeters. To verify this, the analysis is different according to whether the circuit is star- or delta-connnected.

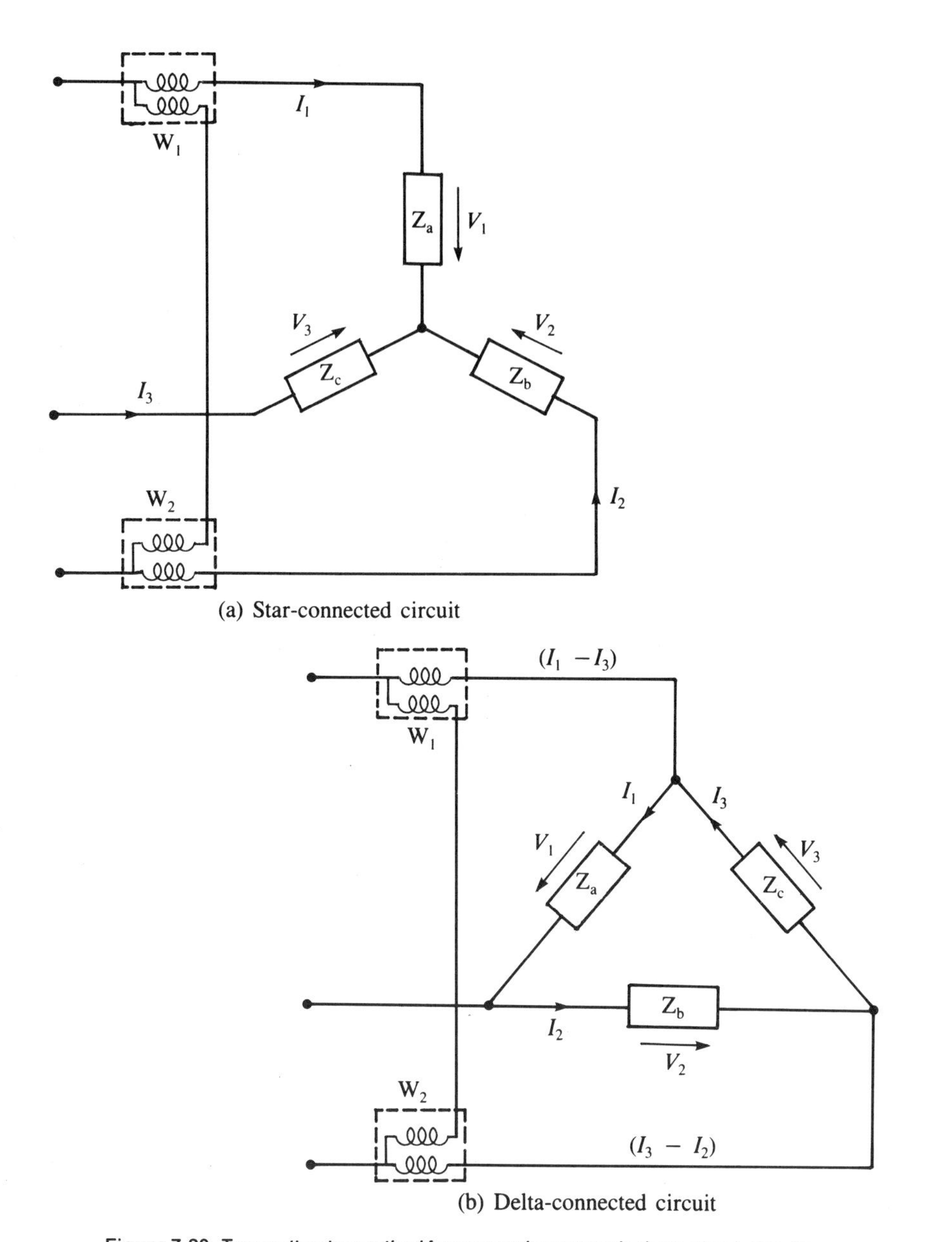

(a) Star-connected circuit

(b) Delta-connected circuit

Figure 7.20 *Two wattmeter method for measuring power in three phase circuits*

For a star-connected circuit, referring to Figure 7.20(a): net current at neutral point = 0, i.e.

$$I_1 + I_2 + I_3 = 0 \tag{7.4}$$

The instantaneous power P is given by:

$$P = V_1 I_1 + V_2 I_2 + V_3 I_3 \tag{7.5}$$

From equation (7.4),

$$I_3 = -(I_1 + I_2)$$

Hence, substituting for I_3 in (7.5):

$$P = V_1 I_1 + V_2 I_2 - V_3(I_1 + I_2) = I_1(V_1 - V_3) + I_2(V_2 - V_3) \tag{7.6}$$

In equation (7.6), I_1 and I_2 are the currents in wattmeters 1 and 2 respectively. Also, $(V_1 - V_3)$ and $(V_2 - V_3)$ are the voltages across wattmeters 1 and 2 respectively. Thus, equation (7.6) can be rewritten as:

$$P = p_1 + p_2 \tag{7.7}$$

where p_1 and p_2 are the instantaneous powers measured by W_1 and W_2 respectively. Because of the inertial properties of wattmeters, the readings p_1 and p_2 are in fact the average powers. Thus the average power P is given by equation (7.7).

For a delta-connected network, referring to Figure 7.20(b):

$$V_1 + V_2 + V_3 = 0 \tag{7.8}$$

The instantaneous power P is given by:

$$P = V_1 I_1 + V_2 I_2 + V_3 I_3 \tag{7.9}$$

From equation (7.8), $V_3 = -(V_1 + V_2)$. Substituting for V_3 in equation (7.9):

$$P = V_1 I_1 + V_2 I_2 - I_3(V_1 + V_2) = V_1(I_1 - I_3) + V_2(I_2 - I_3) \tag{7.10}$$

In equation (7.10), V_1 and V_2 are the voltages across wattmeters 1 and 2 respectively. Also, $(I_1 - I_3)$ and $(I_2 - I_3)$ are the currents in wattmeters 1 and 2 respectively. Hence, equation (7.10) can be rewritten as:

$$P = p_1 + p_2 \tag{7.11}$$

where p_1 and p_2 are the instantaneous powers measured by W_1 and W_2 respectively.

Because of the inertial properties of wattmeters, the readings p_1 and p_2 are in fact the average powers. Thus the average power P is given by equation (7.11).

7.8 Frequency measurement

Frequency measurement is required as part of those instruments that convert the measured physical quantity into a frequency change, such as the variable-reluctance velocity transducer, stroboscopes, the vibrating-wire force sensor, resonant-wire pressure measuring instruments and the quartz thermometer. The output relationship of some forms of a.c. bridge circuit used for measuring inductance and capacitance also requires the excitation frequency to be known.

The instrument that is now most widely used for measuring frequency, which is measured in units of *hertz* (Hz), is the digital counter-timer. The oscilloscope is also commonly used for obtaining approximate measurements of frequency, especially in circuit test and fault-diagnosis applications. Within the audio frequency range, the Wien bridge is a further instrument which is often used. Certain other frequency measuring instruments also exist which are used in some special applications, including the dynamometer frequency meter, the zero-beat meter, the grid-dip meter, Campbell's circuit and Maxwell's commutator bridge. These are discussed in detail elsewhere (Baldwin, 1973; Wolf, 1973).

7.8.1 **Digital counter-timers**

Digital counter-timers are the most accurate and flexible instrument available for measuring frequency. An inaccuracy of 1 part in 10^8 can be achieved, and all frequencies between d.c. and a few gigahertz can be measured.

The essential component within a counter-timer instrument is an oscillator which provides a very accurately known and stable reference frequency, which is typically either 100 kHz or 1 MHz. This is often maintained in a temperature-regulated environment within the instrument to guarantee its accuracy. The oscillator output is transformed by a pulse-shaper circuit into a train of pulses and applied to an electronic gate, as shown in Figure 7.21. Successive pulses at the reference frequency alternately open and close the gate. The input signal of unknown frequency is similarly transformed into a train of pulses and applied to the gate. The number of these pulses which get through the gate during the time that it is open during each gate cycle is proportional to the frequency of the unknown signal.

The accuracy of measurement obviously depends upon how far the unknown frequency is above the reference frequency. As it stands, therefore, the instrument can only accurately measure frequencies which are substantially above 1 MHz. To enable the instrument to measure much lower frequencies, a series of decade frequency-dividers are provided within it. These increase the time between the reference

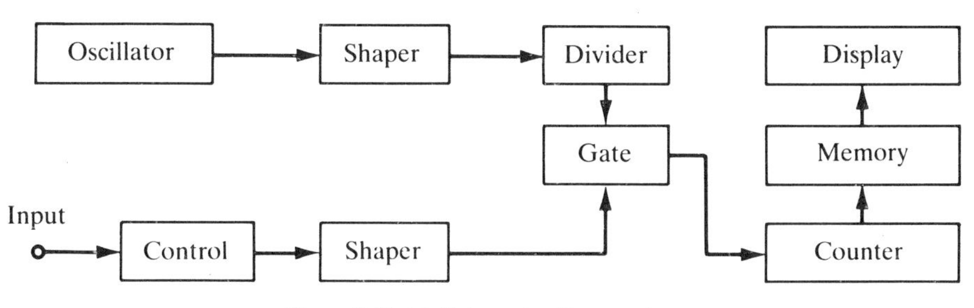

Figure 7.21 *Digital counter-timer system*

frequency pulses by factors of ten, and a typical instrument can have gate pulses separated in time by between 1 μs and 1 s.

Improvement in the accuracy of low-frequency measurement can be obtained by modifying the gating arrangements such that the signal of unknown frequency is made to control the opening and closing of the gate. The number of pulses at the reference frequency which pass through the gate during the open period is then a measure of the frequency of the unknown signal.

7.8.2 **Phase-locked loop**

A phase-locked loop is a circuit consisting of a phase-sensitive detector, a low-pass filter and a voltage controlled oscillator (VCO), connected in a closed-loop system as shown in Figure 7.22. In a VCO, the oscillation frequency is proportional to the applied control voltage.

Operation of a phase-locked loop is as follows. The phase-sensitive detector compares the phase of the input signal with the phase of the VCO output. Any phase difference generates a d.c. error signal, which is amplified and fed back to the VCO. This adjusts the frequency of the VCO until the error signal goes to zero, and thus the VCO becomes locked to the frequency of the input signal. The error signal is used to provide a d.c. output proportional to the input signal frequency.

7.8.3 **Cathode-ray oscilloscope**

The cathode-ray oscilloscope can be used in two ways to measure frequency. Firstly, the internal time-base can be adjusted until the distance between two successive cycles of the measured signal can be read against the calibrated graticule on the screen. Measurement accuracy by this method is limited, but can be optimized by measuring between points in the cycle where the slope of the waveform is steep, generally where

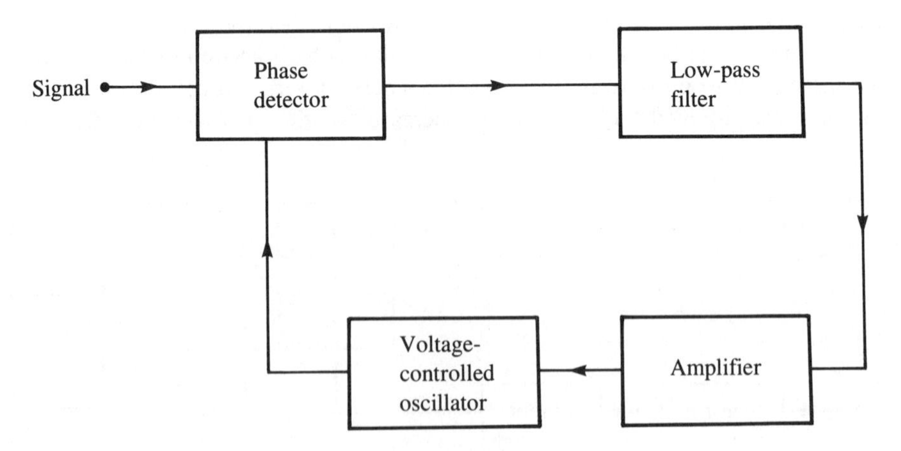

Figure 7.22 *Phase-locked loop*

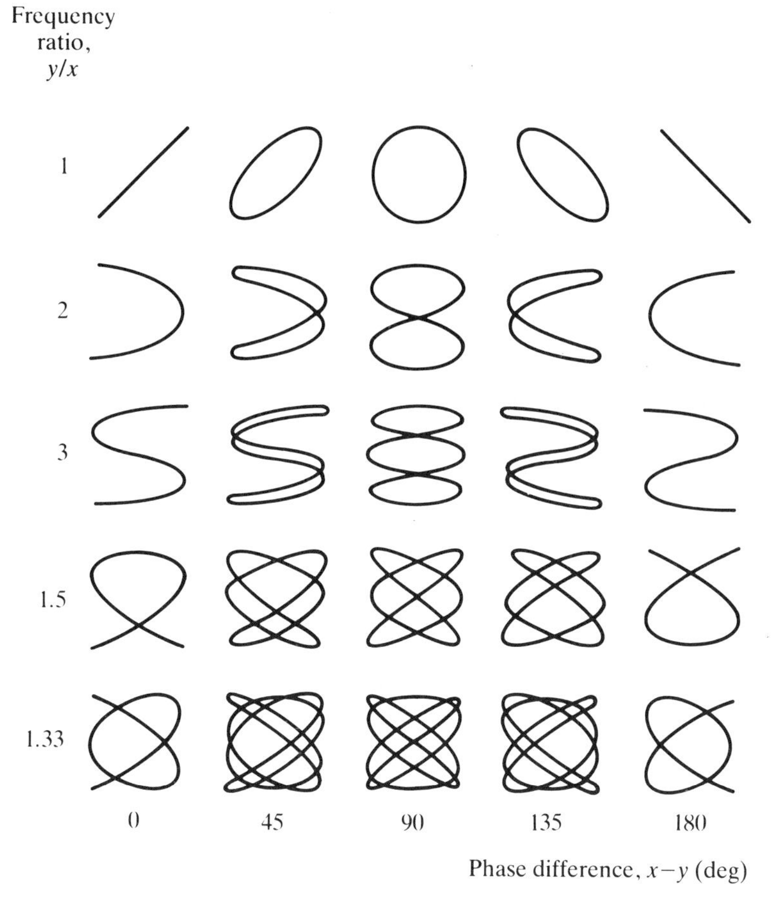

Figure 7.23 *Lissajous patterns*

it is crossing through from the negative to the positive part of the cycle. Calculation of the unknown frequency from this measured time interval is relatively simple. For example, suppose that the distance between two cycles is 2.5 divisions when the internal time-base is set at 10 ms/div. The cycle time is therefore 25 ms and hence the frequency is 1000/25, i.e. 40 Hz. Measurement accuracy is dependent upon how accurately the distance between two cycles is read, and it is very difficult to reduce the error level below ±5% of the reading.

The alternative way of using an oscilloscope to measure frequency is to generate Lissajous patterns. These are produced by applying a known reference-frequency sine wave to the *y* input (vertical deflection plates) of the oscilloscope and the unknown-frequency sinusoidal signal to the *x* input (horizontal deflection plates). A pattern is produced on the screen according to the frequency ratio between the two signals, and if the numerator and denominator in the ratio of the two signals both represent an integral number of cycles, the pattern is stationary. Examples of these patterns are shown in Figure 7.23, which also shows that phase difference between the waveforms has an effect on the shape. Frequency measurement proceeds by adjusting the reference frequency until a steady pattern is obtained on the screen and then calculating the unknown frequency according to the frequency ratio which the pattern obtained represents.

7.8.4 The Wien bridge

The Wien bridge, shown in Figure 7.24, is a special form of a.c. bridge circuit which can be used to measure frequencies in the audio range. An alternative use of the instrument is as a source of audio frequency signals of accurately known frequency.

A simple set of headphones is often used to detect the null-output balance condition. Other suitable instruments for this purpose are the oscilloscope and the electronic voltmeter. At balance, the unknown frequency is calculated according to:

$$f = \frac{1}{2\pi R_3 C_3}$$

The instrument is very accurate at audio frequencies, but at higher frequencies errors due to losses in the capacitors and stray capacitance effects become significant.

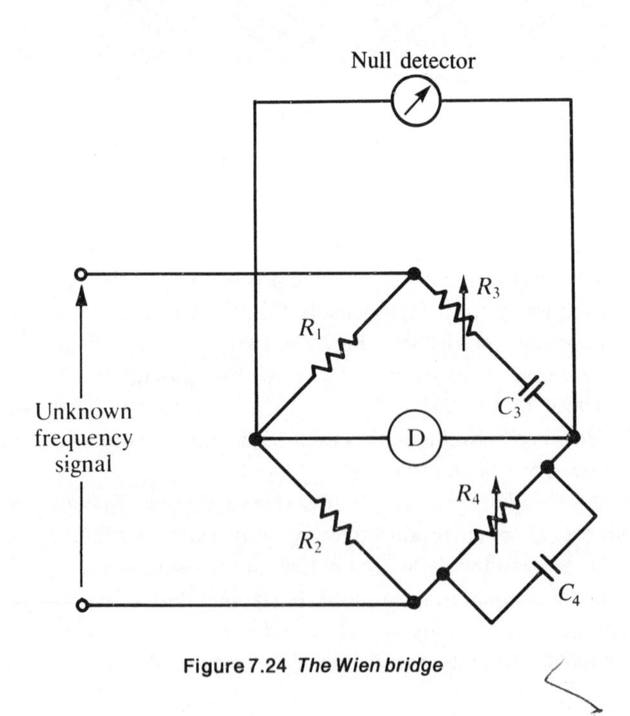

Figure 7.24 *The Wien bridge*

7.9 **Phase measurement**

The most accurate instrument for measuring the phase difference between two signals is the electronic counter-timer. However, two other methods also exist which are less accurate but very useful in some circumstances. These are plotting the signals on an *X–Y* plotter and using a dual beam oscilloscope.

7.9.1 *Electronic counter-timer*

In principle, the phase difference between two sinusoidal signals can be determined by measuring the time which elapses between the two signals crossing the time axis. However, in practice, this is inaccurate because the zero crossings are susceptible to noise contamination. The normal solution to this problem is to amplify/attenuate the two signals so that they have the same amplitude and then measure the time which elapses between the two signals crossing some non-zero threshold value.

The basis of phase measurement by this method is a digital counter-timer with a quartz-controlled oscillator providing a frequency standard which is typically 10 MHz. The crossing points of the two signals through the reference threshold voltage level are applied to a gate which starts and then stops pulses from the oscillator into an electronic counter, as shown in Figure 7.25. The elapsed time, and hence phase difference, between the two input signals is then measured in terms of the counter display.

7.9.2 *X–Y plotter*

This is a useful technique for phase measurement but is limited to low frequencies because of the very limited bandwidth of an *X–Y* plotter. If two sinusoidal input signals of equal magnitude but with a phase difference are applied to the *X* and *Y* inputs of a

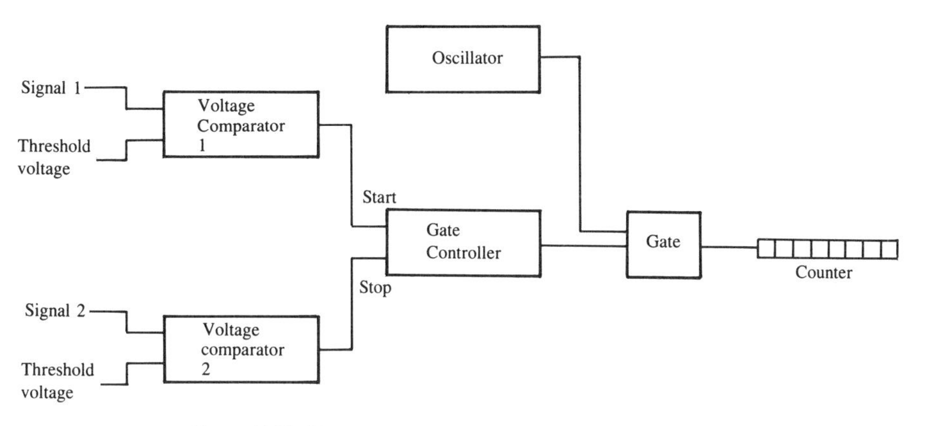

Figure 7.25 *Phase measurement with digital counter-timer*

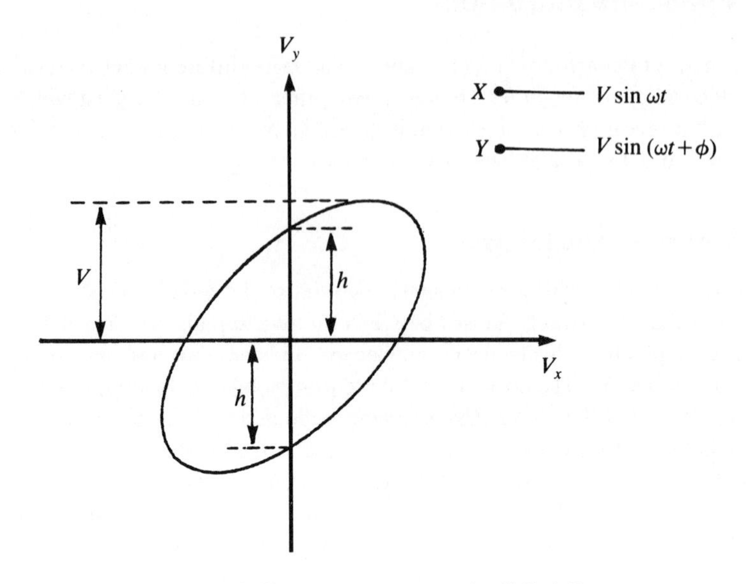

Figure 7.26 *Phase measurement using X–Y plotter*

plotter, the plot obtained is an ellipse, as shown in Figure 7.26. The X and Y inputs can be described by:

$$V_x = V \sin \omega t; \quad V_y = V \sin(\omega t + \phi)$$

where ω is the supply frequency in rad/s and ϕ is the phase difference in radians.

Then at $t = 0$, $V_x = 0$ and $V_y = V \sin \phi$. From Figure 7.26, for $V_x = 0$, $V_y = \pm h$, hence:

$$\sin \phi = \pm h/V \tag{7.12}$$

Solution of equation (7.12) gives four possible values for ϕ but the ambiguity about which quadrant ϕ is in can usually be solved by observing the two signals plotted against time on a dual-beam oscilloscope.

7.9.3 **Oscilloscope**

Approximate measurement of the phase difference between signals can be made using a dual-beam oscilloscope. The two signals are applied to the two oscilloscope vertical deflection inputs and a suitable time-base chosen such that the time between the crossing points of the two signals can be measured. The phase-difference of both low- and high-frequency signals can be measured by this method, the upper frequency limit measurable being dictated by the bandwidth of the oscilloscope (which is normally very high).

7.10 **Capacitance measurement**

The only methods available for accurately measuring capacitance, which is measured in units of *farads* (F), are the various forms of a.c. bridge described in Chapter 6. Various types of capacitance bridge are available commercially, and these would normally be the recommended way of measuring capacitance. In some circumstances, such instruments are not immediately available, and if an approximate measurement of capacitance is acceptable, one of the following two methods can be considered.

The first of these, shown in Figure 7.27, consists of connecting the unknown capacitor in series with a known resistance in a circuit excited at a known frequency. An a.c. voltmeter is used to measure the potential difference across the resistor and capacitor respectively. The capacitance value is then given by:

$$C = \frac{V_r}{2\pi f R V_c}$$

where V_r and V_c are the voltages measured across the resistance and capacitance respectively, f is the excitation frequency in Hz and R is the known resistance.

An alternative method of measurement is to measure the time-constant of the capacitor connected in an RC-circuit.

7.11 **Inductance measurement**

Inductance is measured in *henry* (H). As in the case of capacitance, it can only be measured accurately by an a.c. bridge circuit. Various commercial inductance bridges are available for this purpose. Such instruments are not always immediately available, however, and the following method can be applied in such circumstances to give an approximate measurement of inductance.

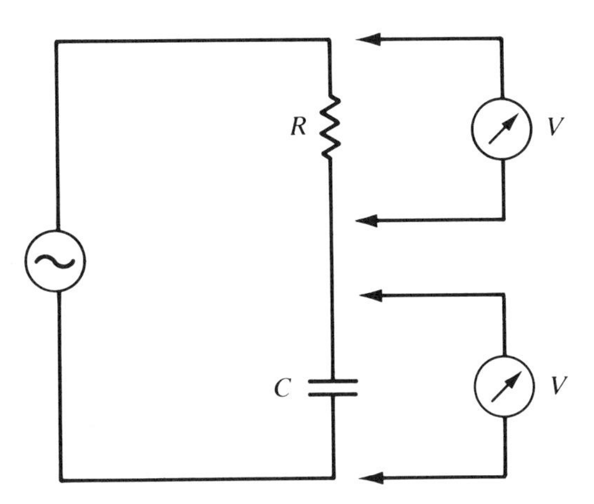

Figure 7.27 *Approximate method of measuring capacitance*

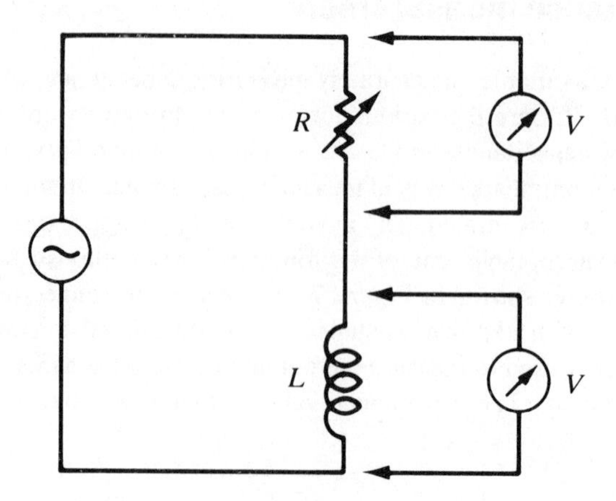

Figure 7.28 *Approximate method of measuring inductance*

This approximate method consists of connecting the unknown inductance in series with a variable resistance, in a circuit excited with a sinusoidal voltage, as shown in Figure 7.28. The variable resistance is adjusted until the voltage measured across the resistance is equal to that measured across the inductance. The two impedances are then equal, and the value of the inductance L can be calculated from:

$$L = \frac{\sqrt{R^2 - r^2}}{2\pi f}$$

where R is the value of the variable resistance, r is the value of the inductor resistance and f is the excitation frequency in Hz.

7.12 Self-assessment questions

7.1 Discuss the dual-ramp technique of analog-to-digital conversion used in some types of digital voltmeter.

7.2 The waveform of a cyclic current is shown in Figure 7.29. The current is measured using (a) a moving-coil ammeter and (b) a moving-iron ammeter. Calculate the readings on the two instruments. (Hint: the current is given by $I = 10 \sin \omega t$ for $0 < \omega t < \pi$, and by $I = 10(1 - \omega t/\pi)$ for $\pi < \omega t < 2\pi$.)

7.3 (a) A moving-iron ammeter and a simple moving-coil ammeter are placed in series with a load in a half-wave rectifier circuit. The reading on the moving-iron instrument is 5 amperes. What is the reading on the other instrument?
(b) Calculate the readings on the two instruments if the other half-wave is also rectified. (Assume sinusoidal waveforms throughout)

7.4 (a) Describe the construction and operating principles of a repulsion-type moving-iron meter.

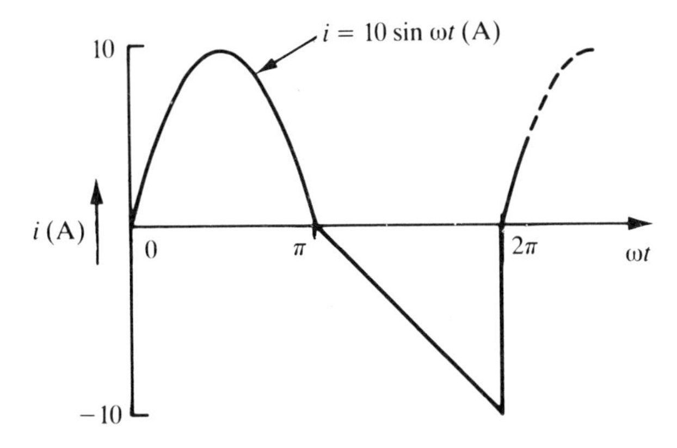

Figure 7.29

(b) A compound voltage consisting of a d.c. level component of 5 V, together with a superimposed half-wave rectified sinusoidal waveform of peak value 4 V, occurs at a certain point in an electrical circuit. What would be the reading on (i) a moving-coil meter and (ii) a moving-iron meter when used to measure this voltage waveform?

7.5 If the waveform shown in Figure 7.30 below was fed to
(a) a true r.m.s. meter,
(b) an average-measuring/r.m.s.-indicating meter,
(c) a peak-measuring/r.m.s.-indicating meter,
what would the reading be on each instrument?

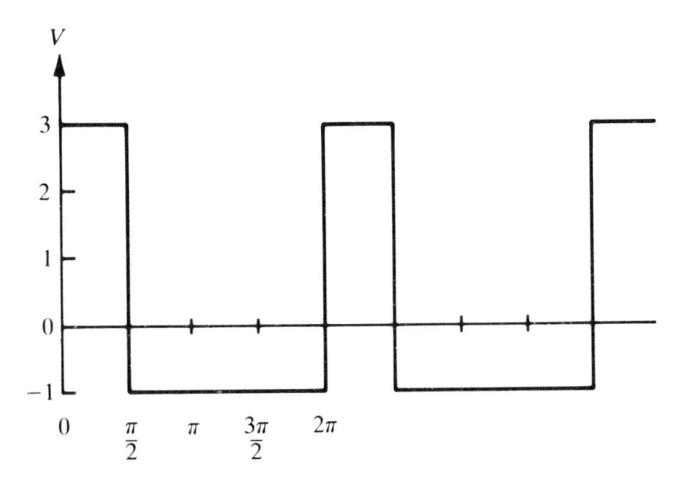

Figure 7.30

7.6 (a) Describe the construction and operating principles of a dynamometer (electrodynamic meter).
(b) A dynamometer ammeter, a d.c. moving-coil ammeter and a 200 Ω resistor are connected in series with a rectifying device across a 240 V r.m.s. alternating sinusoidal supply. The rectifier can be considered to behave as a resistance of 100 Ω to current in one direction and of 1 kΩ to that in the opposite direction. Calculate: (i) the readings on the two ammeters and (ii) the power in the 200 Ω resistor.

References and further reading

Baldwin, C.T., *Fundamentals of Electrical Measurements*, 1973. Prentice Hall, London.
Buckingham, H. and Price, E.M., *Principles of Electrical Measurements*, 1966. English Universities Press, London.
Edwards, D.F.A., *Electronic Measurement Techniques*, 1971. Butterworths, London.
Harris, F.K., *Electrical Measurements*, 1952. Wiley, New York.
Smith, R.J., *Circuits, Devices and Systems*, 1976. Wiley, New York.
Wolf, S., *Guide to Electronics Measurements and Laboratory Practice*, 1973. Prentice Hall, Englewood Cliffs, New Jersey.

CHAPTER 8

Data recording and presentation

8.1 Introduction

The earlier chapters in this book have been essentially concerned with describing ways of producing high-quality, error-free data at the output of a measurement system. Having got the data, the next consideration is how to present it in a form where it can be readily used and analyzed. This generally means writing it down or displaying it in some other way on paper. However, as an intermediate step, there is frequently a requirement to record the data in some other, non-paper form, from where it can be transferred to paper later.

This chapter therefore starts with a discussion of the various means of recording data, and includes chart recorders, ultraviolet recorders, fiber-optic recorders, magnetic tape recorders and computer data logging. Following this, the relative merits of presenting data on paper in tabular and graphical forms are compared. This leads on to a discussion of mathematical regression techniques for fitting the best lines through data points on a graph. Confidence tests to assess the correctness of the line fitted are also described. Finally, correlation tests are described which determine the degree of association between two sets of data when they are both subject to random fluctuations.

8.2 Recording of data

Chart recorders are probably the most widely used method of recording data and carry out this task simply, cheaply and reliably. However, their dynamic response is poor and they cannot record signals with frequencies greater than about 30 Hz. Ultraviolet recorders can cope with much higher frequencies up to about 13 kHz, but they are very delicate and easily damaged instruments which are also quite expensive. Fiber-optic recorders are even more expensive than ultraviolet recorders but have a higher bandwidth still. These different types of recorder, together with magnetic tape and computer data logging, are discussed below.

8.2.1 Chart recorders

Chart recorders provide a relatively simple and cheap way of making permanent

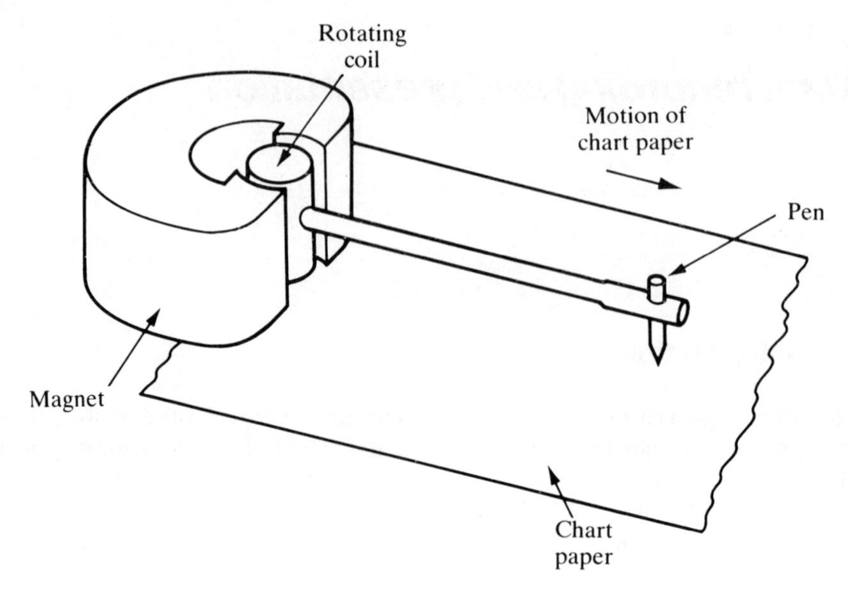

Figure 8.1 *Simple galvanometric recorder*

records of electrical signals. The two main kinds of chart recorder are the galvanometric type and the potentiometric type, with the former accounting for about 80% of the market.

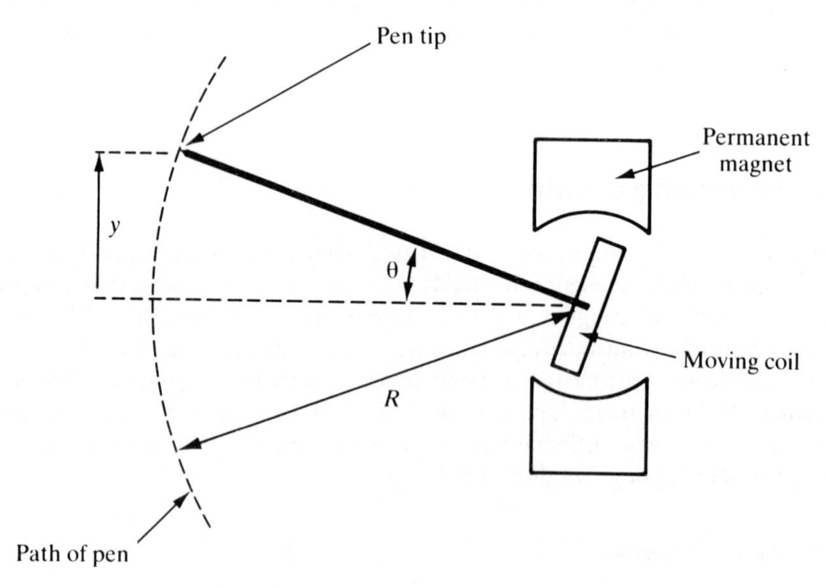

Figure 8.2 *y vs. θ relationship for standard chart recorder*

Galvanometric recorders

These work on the same principle as a moving-coil meter except that the pointer draws an ink trace on paper, as illustrated in Figure 8.1, instead of merely moving against a scale. The measured signal is applied to the coil, and the angular deflection of this and its attached pointer is proportional to the magnitude of the signal applied. The pointer normally carries a pen, and, by using a motor running at constant speed to drive the chart paper which the pointer is in contact with, a time history of the measured signal is obtained.

Inspection of Figure 8.2 shows that the displacement y of the pen across the chart recorder is given by $y = R \cdot \sin \theta$. This sine relationship between the input signal and the displacement y is non-linear, and results in an error of 0.7% for deflections of $\pm 10°$. A more serious problem resulting from the pen moving in an arc is that it is difficult to relate the magnitude of deflection with the time axis. One way of overcoming this is to print a grid on the chart paper in the form of circular arcs, as illustrated in Figure 8.3.

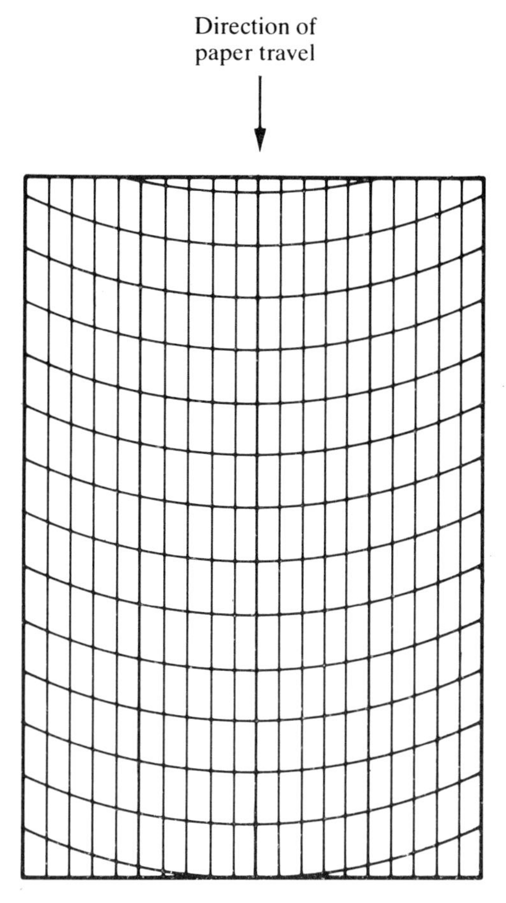

Figure 8.3 *Curvilinear paper for standard galvanometric recorder*

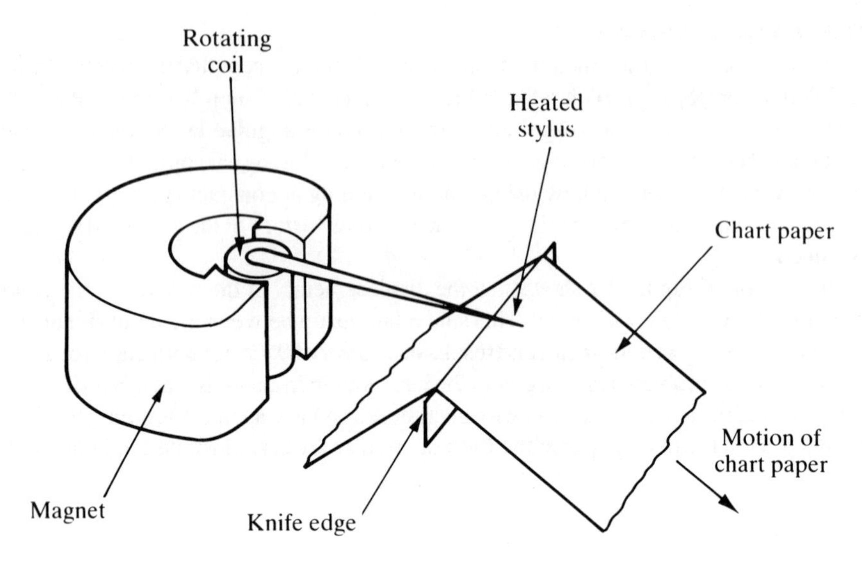

Figure 8.4 ***Knife-edge galvanometric recorder***

Unfortunately, measurement errors often occur in reading this type of chart, as interpolation for points drawn between the curved grid lines is difficult. An alternative solution is to use heat-sensitive chart paper directed over a knife edge, and to replace the pen by a heated stylus, as illustrated in Figure 8.4. The input–output relationship is still non-linear, with the deflection y being proportional to $\tan\theta$ as shown in Figure 8.5,

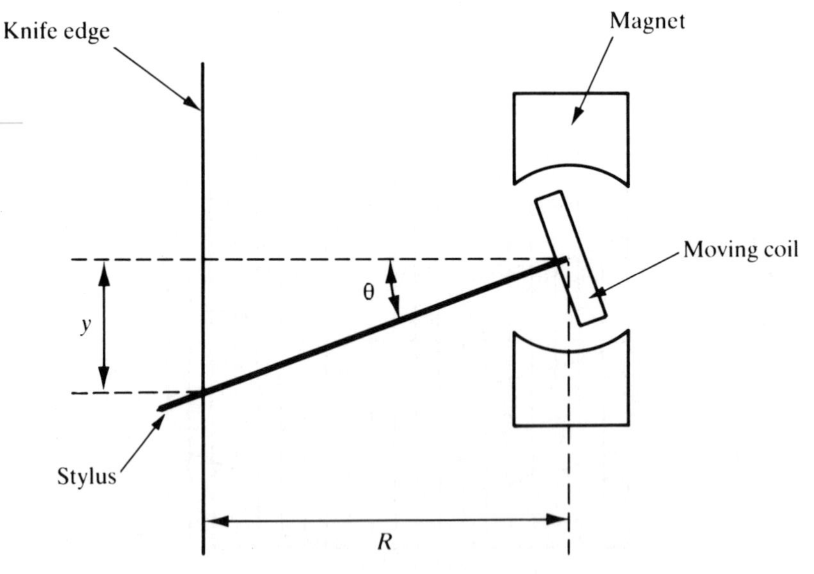

Figure 8.5 ***y vs. θ relationship for knife-edge recorder***

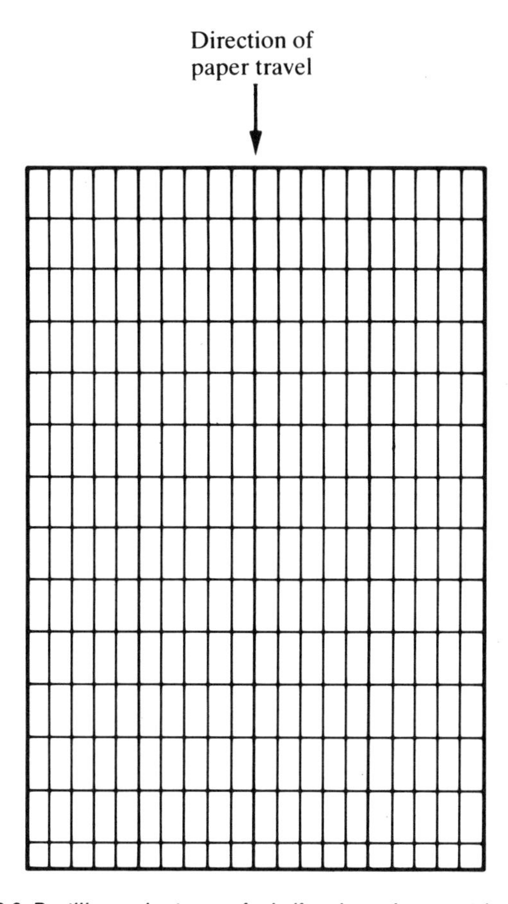

Figure 8.6 *Rectilinear chart paper for knife-edge galvanometric recorder*

and the reading error for excursions of ±10° is still 0.7%. However, the rectilinearly scaled chart paper now required, as shown in Figure 8.6, allows much easier interpolation between grid lines.

Galvanometric recorders have a typical inaccuracy of ±2% and a resolution of 1%. Measurement with galvanometric chart recorders is limited to signals below about 30 Hz in frequency, the reason for which is readily understood if their dynamic behaviour is examined.

Dynamic behaviour of galvanometric recorders

Neglecting friction, the torque equation with the recorder in steady state can be expressed as:

torque due to current in coil = torque due to spring

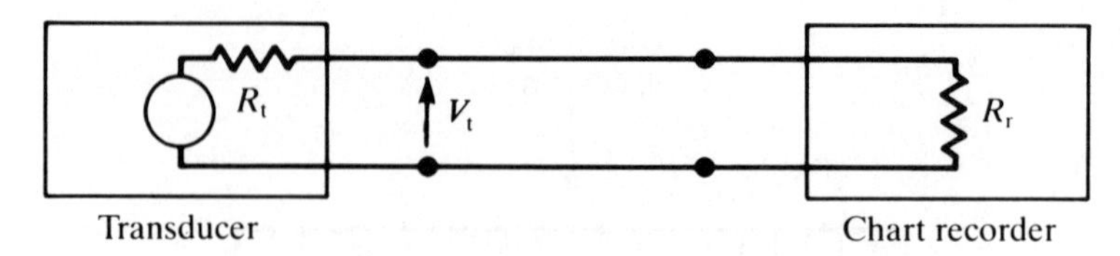

Figure 8.7 *Connection of transducer to chart recorder*

Following a step input, we can write:

torque due to current in coil = torque due to spring + accelerating torque

or:

$$K_i \cdot I = K_s \cdot \theta + J \cdot \ddot{\theta} \tag{8.1}$$

where I is the coil current, θ is the angular coil displacement, $\ddot{\theta}$ represents $d^2\theta/dt^2$, the angular coil acceleration, J is the moment of inertia of the moving components and K_i and K_s are constants.

Consider now what happens if a recorder with resistance R_r is connected to a transducer with resistance R_t and output voltage V_t, as shown in Figure 8.7. The current I in the steady state is given by:

$$I = V_t/(R_t + R_r)$$

When the transducer voltage V_t is first applied to the recorder coil, the coil will accelerate and, because the coil is moving in a magnetic field, a backwards voltage will be induced in it given by

$$V_i = -K_i \cdot \dot{\theta}$$

where $\dot{\theta}$ represents $d\theta/dt$, the angular coil velocity. Hence the coil current is now given by:

$$I = \frac{Vt - K_i \cdot \dot{\theta}}{R_t + R_r}$$

Now substituting for I in the system equation (8.1):

$$K_i\left(\frac{V_t - K_i \cdot \dot{\theta}}{R_t + R_r}\right) = K_s \cdot \theta + J \cdot \ddot{\theta}$$

or rearranging:

$$\ddot{\theta} + \frac{K_i^2 \cdot \dot{\theta}}{J(R_t + R_r)} + \frac{K_s \cdot \theta}{J} = \frac{K_i \cdot V_t}{J(R_t + R_r)} \tag{8.2}$$

This is the equation of a second order dynamic system, with natural undamped frequency ω_n and damping ratio β given by:

$$\omega_n = \sqrt{\frac{K_s}{J}} \qquad \beta = \frac{K_i^2}{2(R_t + R_r)\sqrt{K_s \cdot J}}$$

In steady-state, $\ddot{\theta} = \dot{\theta} = 0$ and equation (8.2) reduces to:

$$\frac{\theta}{V_t} = \frac{K_i}{K_s \cdot (R_t + R_r)} \tag{8.3}$$

which is an expression describing the static sensitivity of the system.

The typical frequency response of a chart recorder for different values of damping ratio is shown in Figure 8.8. This shows that the best bandwidth is obtained at a damping ratio of 0.7, and this is therefore the target when the instrument is designed. The damping ratio depends not only on the coil and spring constants (K_i and K_s) but also on the total circuit resistance ($R_t + R_r$) . Adding a series or parallel resistance between the transducer and recorder as illustrated in Figure 8.9 respectively reduces or increases the damping ratio. However, consideration of the sensitivity expression of (8.3) shows that any reduction in the damping ratio takes place at the expense of a reduction in measurement sensitivity. Other methods to alter the damping ratio are therefore usually necessary and these techniques include decreasing the spring

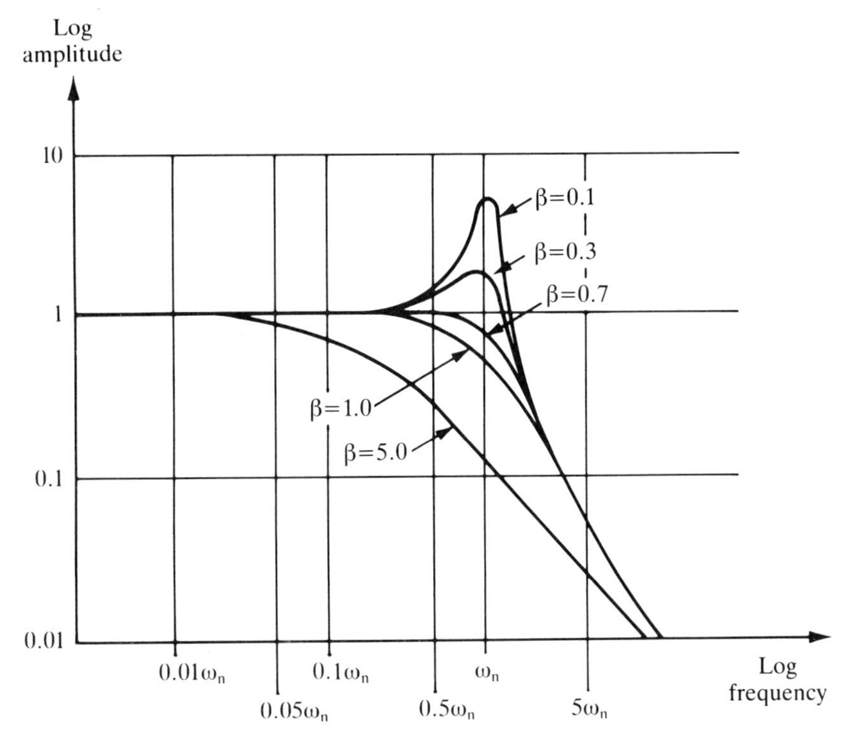

Figure 8.8 *Frequency response of galvanometric chart recorder*

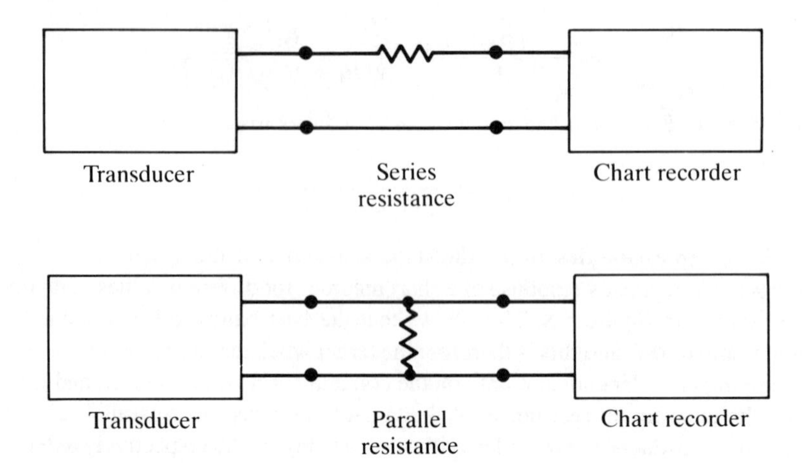

Figure 8.9 *Addition of series and parallel resistances between transducer and chart recorder*

constant and system moment of inertia. In practice, there is a limit as to how far the dynamic performance of a chart recorder can be improved and other instruments such as ultraviolet recorders have to be used for higher frequency signals.

Potentiometric recorders

Potentiometric recorders can achieve an inaccuracy of ±0.1% or less of full scale and measurement resolution of 0.2% f.s., which are both much better than the corresponding figures for galvanometric recorders. They employ a servo system, as shown in Figure 8.10, in which the pen is driven by a servomotor, and a potentiometer on the pen feeds back a signal proportional to pen position. This position signal is compared with the measured signal, and the difference is applied as an error signal which drives the motor. However, a consequence of this electromechanical balancing

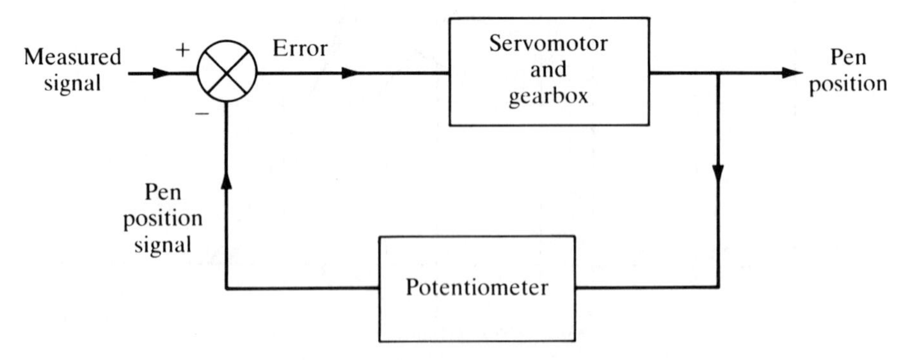

Figure 8.10 *Servo system of potentiometer recorder*

mechanism is to give the instrument a slow response time in the range of 0.2–2.0 seconds. This means that potentiometric recorders are only suitable for measuring d.c. and slowly time-varying signals.

8.2.2 *Ultra-violet recorders*

The earlier discussion about galvanometric recorders concluded that restrictions on how far the system moment of inertia and spring constants can be reduced limited the maximum bandwidth to about 100 Hz. Ultraviolet recorders work on very similar principles to standard galvanometric chart recorders, but achieve a very significant reduction in system inertia and spring constants by mounting a narrow mirror rather that a pen system on the moving coil. This mirror reflects a beam of ultraviolet light onto ultraviolet sensitive paper. It is usual to find several of these mirror-galvanometer systems mounted in parallel within one instrument to provide a multi-channel recording capability, as illustrated in Figure 8.11. This arrangement enables signals at frequencies up to 13 kHz to be recorded with a typical inaccuracy of ±2% f.s. While it is possible to obtain satisfactory permanent signal recordings by this method, special precautions are necessary to protect the UV-sensitive paper from light before use and to spray a fixing lacquer on it after recording. Such instruments must also be handled

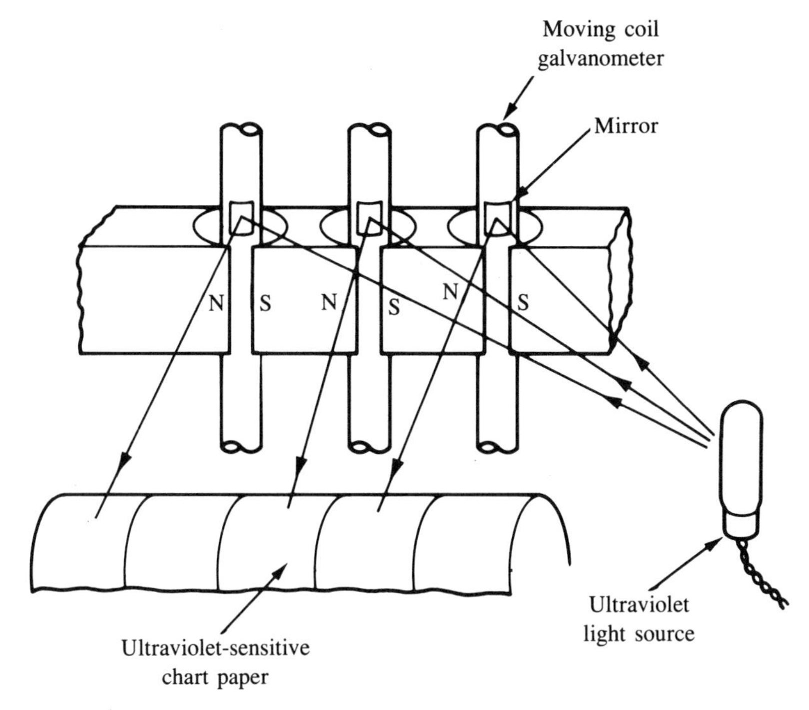

Figure 8.11 *Ultraviolet recorder*

with extreme care, because the mirror galvanometers and their delicate mounting systems are easily damaged by relatively small shocks.

8.2.3 *Fiber-optic recorders*

The fiber-optic recorder is a very recent development which uses a fiber-optic system to direct light onto light sensitive paper. Fiber-optic recorders are similar to oscilloscopes in construction, insofar as they have an electron gun and focusing system which directs a stream of electrons to one point on a fluorescent screen. The screen is usually a long thin one instead of the square type found in an oscilloscope and only one set of deflection plates is provided. The signal to be recorded is applied to the deflection plates and the movement of the focused spot of electrons on the screen is proportional to the signal amplitude. A narrow strip of fiber optics in contact with the fluorescent screen transmits the motion of the spot to photosensitive paper held in close proximity to the other end of the fiber-optic strip. By driving the photosensitive paper at a constant speed past the fiber-optic strip, a time history of the measured signal is obtained. Such recorders are much more expensive than ultraviolet recorders but have an even higher bandwidth.

While the construction above is the more common in fiber-optic recorders, a second type also exists which uses a conventional square screen instead of a long thin one. This has a square faceplate attached to the screen housing a square array of fiber optics. The other side of the fiber-optic system is in contact with chart paper. The effect of this is to provide a hard copy of the typical form of display obtainable on a cathode-ray oscilloscope.

8.2.4 *Magnetic tape recorders*

Magnetic tape recorders can record analog signals up to 80 kHz in frequency. As the speed of the tape transport can be switched between several values, signals can be recorded at high speed and replayed at a lower speed. Such time scaling of the recorded information allows a hard copy of the signal behaviour to be obtained from instruments such as ultraviolet and galvanometric recorders whose bandwidth is insufficient to allow direct signal recording. A 200 Hz signal cannot be recorded directly on a chart recorder, but if it is recorded on a magnetic tape recorder running at high speed and then replayed at a speed ten times lower, its frequency will be time-scaled to 20 Hz which can be recorded on a chart recorder. Instrumentation tape recorders typically have between four and ten channels, allowing many signals to be recorded simultaneously.

The two basic types of analog tape recording technique are direct recording and frequency-modulated recording. Direct recording offers the best data bandwidth but the accuracy of signal amplitude recording is quite poor, and this seriously limits the usefulness of this technique in most applications. The frequency-modulated technique offers much better amplitude recording accuracy, with an inaccuracy of only ±5% at signal frequencies of 80 kHz. In consequence, this technique is very much more common than direct recording.

8.2.5 *Computer data logging*

The only technique available for recording signals at frequencies higher than 80 kHz involves using a digital computer. As the signals to be recorded are usually in analog form, a prerequisite for this is an analog-to-digital converter board within the computer to sample the analog signals and convert them to the digital form in which the computer operates. Careful choice of the sampling interval is necessary so that an accurate digital record of the signal is obtained as explained in Chapter 5, so that problems of aliasing, etc., are not encountered. Some prior analog signal conditioning may also be required in some circumstances when a computer is used for data logging, again as mentioned in Chapter 5.

8.3 Presentation of data

The two formats available for presenting data on paper are tabular and graphical ones and the relative merits of these are compared below. In some circumstances, it is clearly best to use only one of these two alternatives. However, in many data collection exercises, part of the measurements and calculations are expressed in tabular form and part graphically, so making best use of the merits of each technique.

It should be mentioned that a large number of computer packages are now available which present data in tabular and graphical forms, and these are much less laborious than 'paper' presentations produced by human hand.

8.3.1 *Tabular data presentation*

A tabular presentation allows data values to be recorded in a precise way which maintains the accuracy to which the data values were measured. In other words, the data values are written down exactly as measured. Besides recording the raw data values as measured, tables often also contain further values calculated from the raw data.

An example of a tabular data presentation is given in Table 8.1. This records the results of an experiment to determine the strain induced in a bar of material under a range of stresses. Data were obtained by applying a sequence of forces to the end of the bar and using an extensometer to measure the change in length. Values of the stress and strain in the bar are calculated from these measurements and are also included in the table. The final row, which is of crucial importance in any tabular presentation, is the estimate of possible error in each calculated result.

A table of measurements and calculations should conform to several rules as illustrated in Table 8.1:

Table 8.1 *Sample tabular presentation of data*

Table of measured applied forces and extensometer readings and calculations of stress and strain

	Force applied (kN)	*Extensometer reading (divisions)*	*Stress (N/m^2)*	*Strain*
	0	0	0	0
	2	4.0	15.5	19.8×10^{-5}
	4	5.8	31.0	28.6×10^{-5}
	6	7.4	46.5	36.6×10^{-5}
	8	9.0	62.0	44.4×10^{-5}
	10	10.6	77.5	52.4×10^{-5}
	12	12.2	93.0	60.2×10^{-5}
	14	13.7	108.5	67.6×10^{-5}
Possible error in measurements (%)	±0.2	±0.2	±1.5	$\pm 1.0 \times 10^{-5}$

1 The table should have a title which explains what data are being presented within the table.
2 Each column of figures in the table should refer to the measurements or calculations associated with one quantity only.
3 Each column of figures should be headed by a title which identifies the data values contained in the column.
4 The units in which quantities in each column are measured should be stated at the top of the column.
5 All headings and columns should be separated by horizontal (and sometimes vertical) lines.
6 The errors associated with each data value quoted in the table should be given. The form shown in Table 8.1 is a suitable way to do this when the error level is the same for all data values in a particular column. However, if error levels vary, then it is preferable to write the error boundaries alongside each entry in the table.

8.3.2 **Graphical presentation of data**

Presentation of data in graphical form involves some compromise in the accuracy to which the data are recorded, as the exact values of measurements are lost. However, graphical presentation has two important advantages over tabular presentation:

1 Graphs provide a pictorial representation of results which is more readily comprehended than a set of tabular results.
2 Graphs are particularly useful for expressing the quantitative significance of results and showing whether a linear relationship exists between two variables.

Figure 8.12 shows a graph drawn from the stress and strain values given in the Table 8.1. Construction of the graph involves first of all marking the points corresponding to

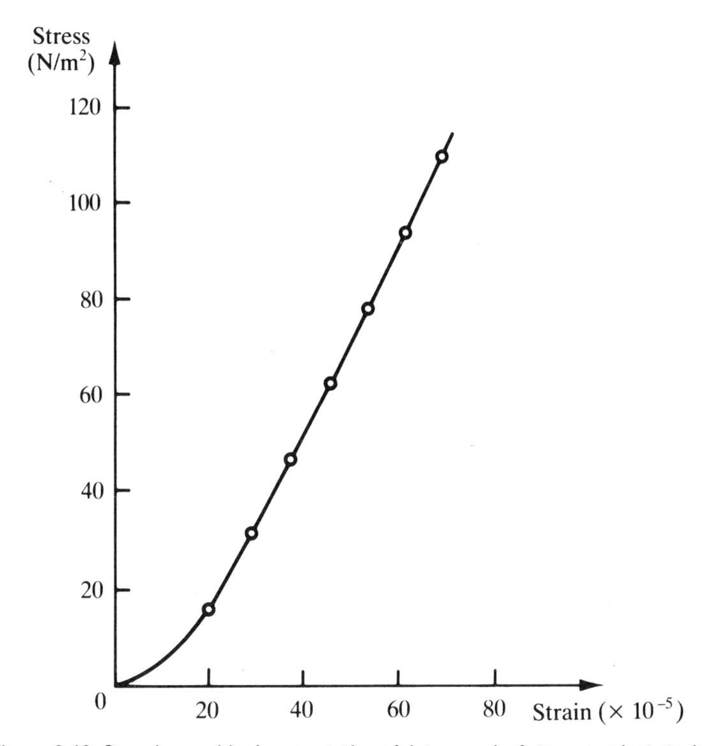

Figure 8.12 *Sample graphical presentation of data: graph of stress against strain*

the stress and strain values. The next step is to draw some line through these data points which best represents the relationship between the two variables. This line will normally be either straight or a smooth curve. The data points will not usually lie exactly on this line but instead will lie on either side of it. The magnitude of the excursions of the data points from the line drawn will depend on the magnitude of the random measurement errors associated with the data.

As for tables, certain rules exist for the proper representation of data in graphical form:

1 The graph should have a title or caption which explains what data are being presented in the graph.
2 Both axes of the graph should be labelled to express clearly what variable is associated with each axis and to define the units in which the variables are expressed.
3 The number of points marked along each axis should be kept reasonably small — about five divisions is often a suitable number.
4 No attempt should be made to draw the graph outside the boundaries corresponding to the maximum and minimum data values measured, i.e. in Figure 8.12, the graph stops at a point corresponding to the highest measured stress value of 108.5.

8.3.3 *Fitting curves to data points on a graph*

The procedure of drawing a straight line or smooth curve as appropriate which passes close to all data points on a graph, rather than joining the data points by a jagged line which passes through each data point, is justified on account of the random errors which are known to affect measurements. Any line between the data points is mathematically acceptable as a graphical representation of the data if the maximum deviation of any data point from the line is within the boundaries of the identified level of possible measurement errors. However, within the range of possible lines which could be drawn, only one will be optimum. This optimum line is where the sum of negative errors in data points on one side of the line is balanced by the sum of positive errors in data points on the other side of the line. The nature of the data points is often such that a perfectly acceptable approximation to the optimum can be obtained by drawing a line through the data points by eye. In other cases, however, it is necessary to fit a line by regression techniques.

8.3.4 *Regression techniques*

Regression techniques consist of finding a mathematical relationship between measurements of two variables y and x, such that the value of one variable y can be predicted from a measurement of the other variable x. It is assumed for this that random errors only affect the y values and that the values of x are exact. Regression procedures are simplest if a straight-line relationship exists between the variables, which can then be estimated by linear least-squares regression.

In many cases, inspection of the raw data points plotted on a graph shows that a straight-line relationship between the points does not exist. However, knowledge of physical laws governing the data can often suggest a suitable alternative form of relationship between the two sets of variable measurements.

In some cases, the measured variables can be transformed such that a linear relationship is obtained. For example, suppose that two variables y and x are related according to: $y = a \cdot x^c$. A linear relationship from this can be derived as:

$$\log(y) = \log(a) + c \cdot \log(x)$$

Thus if a graph is constructed of $\log(y)$ plotted against $\log(x)$, the parameters of a straight-line relationship can be estimated by linear least-squares regression.

Apart from linear relationships, physical laws often suggest a relationship where one variable y is related to another variable x by a power series of the form:

$$y = a_0 + a_1 \cdot x + a_2 \cdot x^2 + \ldots + a_p \cdot x^p$$

Estimation of the parameters $a_0 \ldots a_p$ is very difficult if p has a large value. Fortunately, a relationship where p only has a small value can be fitted to most data sets. Quadratic least-squares regression is used to estimate parameters where p has a value of two, and for larger values of p, polynomial least-squares regression is used for parameter estimation.

Where the appropriate form of relationship between variables in measurement data sets is not obvious either from visual inspection or from consideration of physical laws, a method which is effectively a trial and error one has to be applied. This consists of estimating the parameters of successively higher-order relationships between y and x until a curve is found which fits the data sufficiently closely. What level of closeness is acceptable is considered in the later section on confidence tests.

Linear least-squares regression

If a linear relationship between y and x exists for a set of n measurements $y_1 \ldots y_n, x_1 \ldots x_n$, then this relationship can be expressed as $y = a + b \cdot x$, where the coefficients a and b are constants. The purpose of least-squares regression is to select the optimum values for a and b such that the line gives the best fit to the measurement data.

The deviation of each point (x_i, y_i) from the line can be expressed as d_i, where $d_i = y_i - (a + b \cdot x_i)$.

The best-fit line is obtained when the sum of the squared deviations, S, is a minimum, i.e. when

$$S = \sum_{i=1}^{n} (d_i^2) = \sum_{i=1}^{n} (y_i - a - b \cdot x_i)^2$$

is a minimum. The minimum can be found by setting the partial derivatives $\partial S/\partial a$ and $\partial S/\partial b$ to zero and solving the resulting two simultaneous (normal) equations:

$$\partial S/\partial a = \Sigma\, 2\,(y_i - a - b \cdot x_i)(-1) = 0 \tag{8.4}$$

$$\partial S/\partial b = \Sigma\, 2\,(y_i - a - b \cdot x_i)(-x_i) = 0 \tag{8.5}$$

The values of a and b at the minimum point can be represented by $\hat{a}$ and $\hat{b}$. These parameters $\hat{a}$ and $\hat{b}$, which are known as the least-squares estimates of a and b, can be calculated as follows.

From equation (8.4):

$$\Sigma y_i = \Sigma \hat{a} + \hat{b} \Sigma x_i = n\hat{a} + \hat{b} \Sigma x_i,$$

and thus:

$$\hat{a} = \frac{\Sigma y_i - \hat{b} \Sigma x_i}{n} \tag{8.6}$$

From equation (8.5):

$$\Sigma (y_i x_i) = \hat{a} \Sigma x_i + \hat{b} \Sigma x_i^2 \tag{8.7}$$

Now substitute for $\hat{a}$ in equation (8.7), using equation (8.6):

$$\Sigma(y_i x_i) = \frac{(\Sigma y_i - \hat{b}\Sigma x_i)}{n}\Sigma x_i + \hat{b}\Sigma x_i{}^2$$

Collecting terms in $\hat{b}$:

$$\hat{b}\left[\Sigma x_i{}^2 - \frac{(\Sigma x_i)^2}{n}\right] = \Sigma(x_i y_i) - \frac{\Sigma x_i \Sigma y_i}{n}$$

Rearranging:

$$\hat{b}[\Sigma x_i{}^2 - n\{(\Sigma x_i/n)\}^2] = \Sigma(x_i y_i) - n\Sigma(x_i/n)\Sigma(y_i/n)$$

This can be expressed as:

$$\hat{b}[\Sigma x_i{}^2 - n x_{\mathrm{m}}{}^2] = \Sigma(x_i y_i) - n x_{\mathrm{m}} y_{\mathrm{m}}$$

using x_{m} and y_{m} to represent the mean values of x and y. Thus:

$$\hat{b} = \frac{\Sigma(x_i y_i) - n x_{\mathrm{m}} y_{\mathrm{m}}}{\Sigma x_i{}^2 - n x_{\mathrm{m}}{}^2} \tag{8.8}$$

and from equation (8.6):

$$\hat{a} = y_{\mathrm{m}} - \hat{b} x_{\mathrm{m}} \tag{8.9}$$

If desired, the last equation for $\hat{b}$ can be expressed in the following alternative form (Chatfield, 1983):

$$\hat{b} = \frac{\Sigma(x_i - x_{\mathrm{m}})(y_i - y_{\mathrm{m}})}{\Sigma(x_i - x_{\mathrm{m}})^2}$$

EXAMPLE 8.1

In an experiment to determine the variation of the specific heat of a substance with temperature, the following table of results was obtained:

Temperature (°C):	40	45	50	55	60	65	70	75	80	85	90	95	100
Specific heat (kJ/kg°C):	1.38	1.43	1.46	1.49	1.56	1.57	1.59	1.64	1.69	1.72	1.78	1.83	1.85

Fit a straight line to this data set using least-squares regression and estimate the specific heat at a temperature of 72 °C.

SOLUTION

Let y represent the specific heat and x represent the temperature. Then a suitable straight line is given by $y = a + b \times x$. We can now proceed to calculate estimates for the coefficients a and b using equations (8.8) and (8.9) above.

The first step is to calculate the mean values of x and y. These are found to be $x_{\mathrm{m}} = 70$ and $y_{\mathrm{m}} = 1.6146$. Next, we need to tabulate $(x_i y_i)$ and $(x_i{}^2)$ for each pair of data values:

i	x_i	y_i	x_iy_i	x_i^2
1	40	1.38	55.20	1 600
2	45	1.43	64.35	2 025
3	50	1.46	73.00	2 500
.	.	.	.	.
.	.	.	.	.
.	.	.	.	.
13	100	1.85	185.00	10 000

Now calculating the quantities required in equation (8.8) from this table:

$$n = 13; \quad \Sigma x_i y_i = 1504.90; \quad \Sigma {x_i}^2 = 68\,250$$

$$n x_{\mathrm{m}} y_{\mathrm{m}} = 1469.286; \quad n {x_{\mathrm{m}}}^2 = 63\,700$$

Thus, substituting the values in (8.8):

$$\hat{b} = \frac{1504.90 - 1469.286}{682\,50 - 63\,700} = 0.007\,827\,2$$

Hence, using (8.9),

$$\hat{a} = 1.6146 - 0.007\,827\,2 \times 70 = 1.066\,696$$

Thus,

$$y = 1.066\,696 + 0.007\,827\,2x$$

Using this equation, at a temperature (x) of 72 °C, the specific heat (y) is 1.63 kJ/kg °C.

Note that in this solution, we have only specified the answer to an accuracy of three figures, which is the same accuracy as the measurements. Any greater number of figures in the answer would be meaningless.

Least-squares regression is often appropriate for situations where a straight-line relationship is not immediately obvious, for example where $y \propto x^2$ or $y \propto \exp(x)$.

EXAMPLE 8.2

From theoretical considerations, it is known that the voltage (V) across a charged capacitor decays with time (t) according to the relationship: $V = K\exp(-t/\tau)$. Estimate values for K and τ if the following values of V and t are measured:

V (volts):	4.3	2.8	1.8	1.2	0.8	0.5	0.3
t (seconds):	0	1	2	3	4	5	6

SOLUTION

If $V = K\exp(-t/\tau)$ then: $\log_e V = \log_e K - t/\tau$. Let $y = \log_e V$; $a = \log_e K$; $b = -1/\tau$; $x = t$. Then $y = a + bx$ which is the equation of a straight line. The best estimates of coefficients a and b are given by (using equations 8.8, 8.9):

$$b = \frac{\Sigma(x_i y_i) - n x_m y_m}{\Sigma x_i{}^2 - n x_m{}^2}; \quad a = y_m - b x_m$$

From the given data, $x_m = 3$ and $y_m = 0.1626$.

Now draw a table giving the quantities required to calculate b from the expression above:

V	$\log_e Vt$ (y_i)	(x_i)	$x_i y_i$	x_i^2
4.3	1.46	0	0	0
2.8	1.03	1	1.030	1
1.8	0.588	2	1.176	4
1.2	0.182	3	0.546	9
0.8	−0.223	4	−0.892	16
0.5	−0.693	5	−3.465	25
0.3	−1.20	6	−7.200	36

Using these values:

$$n = 7; \quad \Sigma x_i y_i = -8.805; \quad \Sigma x_i{}^2 = 91$$

$$n x_m y_m = 3.4146; \quad n x_m{}^2 = 63$$

Thus

$$b = \frac{-8.805 - 3.4146}{91 - 63} = -0.4364$$

$a = 0.1626 + 0.4364 \times 3 = 1.4718$
$K = \exp(a) = \exp(1.4718) = 4.35707$
$\tau = -1/b = (-1)/(-0.4364) = 2.29$

Quadratic least-squares regression

Quadratic least-squares regression is used to estimate the parameters of a relationship $y = a + b \cdot x + c \cdot x^2$ between two sets of measurements $y_1 \ldots y_n$ and $x_1 \ldots x_n$.

The deviation of each point (x_i, y_i) from the line can be expressed as d_i, where $d_i = y_i - (a + b \cdot x_i + c \cdot x_i)^2$. The best fit line is obtained when the sum of the squared deviations, S, is a minimum, i.e. when

$$S = \sum_{i=1}^{n} = \Sigma(d_i^2) = \sum_{i=1}^{n}(y_i - a - b \cdot x_i - c \cdot x_i{}^2)^2$$

is a minimum.

The minimum can be found by setting the partial derivatives $\partial S/\partial a$, $\partial S/\partial b$ and$\partial S/\partial c$ to zero and solving the resulting simultaneous equations, as for the linear least-squares regression case above. Standard computer programs to estimate the parameters a, b and c by numerical methods are widely available and therefore a detailed solution is not presented here.

Polynomial least-squares regression

Polynomial least-squares regression is used to estimate the parameters of the pth order relationship $y = a_0 + a_1 \cdot x + a_2 \cdot x^2 + \ldots a_p \cdot x^p$ between two sets of measurements $y_1 \ldots y_n$ and $x_1 \ldots x_n$.

The deviation of each point (x_i, y_i) from the line can be expressed as d_i, where

$$d_i = y_i - (a_0 + a_1 \cdot x_i + a_2.x_i^2 \ldots + a_p \cdot x_i{}^p)$$

The best fit line is obtained when the sum of the squared deviations, s, is a minimum, i.e. when

$$S = \sum_{i=1}^{n} (d_i{}^2)$$

is a minimum.

The minimum can be found as before by setting the p partial derivatives $\partial S / \partial a_0 \ldots \partial S / \partial a_p$ to zero and solving the resulting simultaneous equations. Again, as for the quadratic least-squares regression case, standard computer programs to estimate the parameters $a_0 \ldots a_p$ by numerical methods are widely available and therefore a detailed solution is not presented here.

8.3.5 *Confidence tests in curve fitting*

Having applied least-squares regression to estimate the parameters of a chosen relationship, some form of follow-up procedure is clearly required to assess how well the estimated relationship fits the data points. One fundamental requirement in curve fitting is that the maximum deviation d_i of any data point (y_i, x_i) from the fitted curve is less than the calculated maximum measurement error level. For some data sets, it is impossible to find a relationship between the data points which satisfies this requirement. This normally occurs when both variables in a measurement data set are subject to random variation, such as where the two sets of data values are measurements of human height and weight. Correlation analysis is applied in such cases to determine the degree of association between the variables.

Assuming that a curve can be fitted to the data points in a measurement set without violating the fundamental requirement that only one, not both, of the measured variables are subject to random errors, a further simple curve-fitting confidence test is to calculate the sum of squared deviations S for the chosen y/x relationship and to compare it with the value of S calculated for the next higher-order regression line which could be fitted to the data. Thus if a straight-line relationship is chosen, the value of S calculated should be of a similar magnitude to that obtained by fitting a quadratic relationship. If the value of S were substantially lower for a quadratic relationship, this would indicate that a quadratic relationship was a better fit to the data than a straight-line one and further tests would be needed to examine whether a cubic or higher-order relationship was a better fit still.

Other, more sophisticated, confidence tests such as the F-ratio test are covered in more advanced texts such as Chatfield (1983).

8.3.6 *Correlation tests*

Where both variables in a measurement data set are subject to random fluctuations, correlation analysis is applied to determine the degree of association between the variables. For example, in the case already quoted of a data set containing measurements of human height and weight, we certainly expect some relationship between the variables of height and weight because a tall person is heavier *on average* than a short person. Correlation tests determine the strength of the relationship (or interdependence) between the measured variables, which is expressed in the form of a correlation coefficient.

For two sets of measurements $x_1 \ldots x_n$ and $y_1 \ldots y_n$ with means x_m and y_m, the correlation coefficient ϕ is given by:

$$\phi = \frac{\Sigma(x_i - x_\mathrm{m})(y_i - y_\mathrm{m})}{\sqrt{[\Sigma(x_i - x_\mathrm{m})^2][\Sigma(y_i - y_\mathrm{m})^2]}}$$

The value of $|\phi|$ always lies between 0 and 1, with 0 representing the case where the variables are completely independent of one another and 1 the case where they are totally related to one another.

For $0 < |\phi| < 1$, linear least-squares regression can be applied to find relationships between the variables, which allows x to be predicted from a measurement of y, and y to be predicted from a measurement of x. This involves finding two separate regression lines of the form: $y = a + b \cdot x$ and $x = c + d \cdot y$. These two lines are not normally coincident as shown in Figure 8.13. Both lines pass through the centroid of the data points but their slopes are different.

As $|\phi| \to 1$, the lines tend to coincidence, representing the case where the two variables are totally dependent upon one another.

As $|\phi| \to 0$, the lines tend to orthogonal ones parallel to the x and y axes. In this case, the two sets of variables are totally independent. The best estimate of x given any measurement of y is x_m and the best estimate of y given any measurement of x is y_m.

For the general case, the best fit to the data is that line which bisects the angle between the lines on Figure 8.13.

8.4 Self-assessment questions

8.1 (a) Derive from first principles the expression:

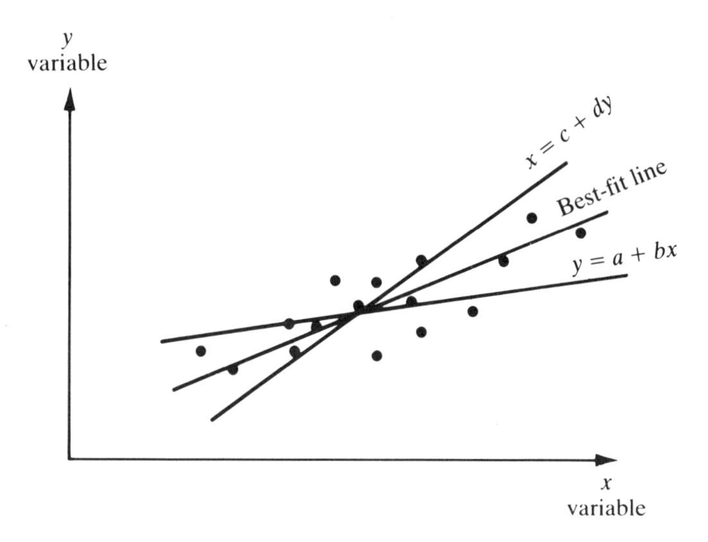

Figure 8.13 *Relationship between two variables with random fluctuations*

$$\ddot{\theta} + \frac{K_I{}^2\dot{\theta}}{JR} + \frac{K_s\theta}{J} = \frac{K_I V_t}{JR}$$

describing the dynamic response of a chart recorder following a step change in the electrical voltage output of a transducer connected to its input. Explain also what all the terms in the expression stand for. (Assume that the impedances of both the transducer and recorder have a resistive component only and that there is negligible friction in the system.)

(b) Derive expressions for the measuring system natural undamped frequency, ω_n, the damping ratio, β, and the steady-state sensitivity.

(c) Explain simple ways of increasing and decreasing the damping ratio and describe the corresponding effect on measurement sensitivity.

(d) What damping ratio gives the best system bandwidth?

(e) What aspects of the design of a chart recorder would you modify in order to improve the system bandwidth? What is the maximum-bandwidth typically attainable in chart recorders, and if such a maximum-bandwidth instrument is available, what is the highest-frequency signal that such an instrument would be generally regarded as being suitable for measuring if the accuracy of the signal amplitude measurement is important?

8.2 The characteristics of a chromel–constantan thermocouple is known to be approximately linear over the temperature range of 300–800 °C. The output e.m.f. (E) was measured practically at a range of temperatures (T) and the following table of results obtained. Using least-squares regression, calculate the coefficients a and b for the relationship $T = a \cdot E + b$ which best describes the temperature–e.m.f. characteristic.

Temp (°C)	300	325	350	375	400	425	450	475	500	525	550
e.m.f. (mV)	21.0	23.2	25.0	26.9	28.6	31.3	32.8	35.0	37.2	38.5	40.7

Temp (°C)	575	600	625	650	675	700	725	750	775	800
e.m.f. (mV)	43.0	45.2	47.6	49.5	51.1	53.0	55.5	57.2	59.0	61.0

8.3 Measurements of the current (I) in resistor and the corresponding voltage drop (V) are shown below:

I (amperes)	1	2	3	4	5
V (volts)	10.8	20.4	30.7	40.5	50.0

The instruments used to measure voltage and current were accurate in all respects except that they each had a zero error which the observer failed to take account of or to correct at the time of measurement. Determine the value of the resistor from the data measured.

8.4 A certain measured quantity y is known from theoretical considerations to depend on a certain variable x according to the relationship:

$$y = a + bx^2$$

For the following set of measurements of x and y, use linear least-squares regression to determine the estimates of the parameters a and b which fit the data best.

x:	0	1	2	3	4	5
y:	0.9	9.2	33.4	72.5	130.1	200.8

8.5 The mean-time-to-failure (MTTF) of an integrated circuit is known to obey a law of the following form:

$$\text{MTTF} = C \exp(T_0/T)$$

where T is the operating temperature and C and T_0 are constants. The following values of MTTF at various temperatures were obtained from accelerated-life tests:

MTTF (hours):	54	105	206	411	941	2145
Temperature (K):	600	580	560	540	520	500

(a) Estimate the values of C and T_0. (Hint: $\log_e(\text{MTTF}) = \log_e(C) + T_0/T$. This equation is now a straight-line relationship between log(MTTF) and $1/T$, where $\log(C)$ and T_0 are constants.)
(b) For a MTTF of 10 years, calculate the maximum allowable temperature.

References and further reading

Chatfield, C., *Statistics for Technology*, 1983. Chapman and Hall, London.
Topping, J., *Errors of Observation and their Treatment*, 1960. Chapman and Hall, London.

CHAPTER 9

Common transducers

The purpose of this final chapter is to give the reader a brief introduction to some transducers and instruments which are commonly used for measuring frequently met physical quantities. The coverage is necessarily brief, but other texts can be consulted if greater information is required (e.g. Morris, 1993).

9.1 Temperature measurement

Temperature measurement is very important in all spheres of life and especially so in the process industries. The main instruments used to measure temperature can be divided into four separate classes according to the physical principle on which they operate. These principles are as follows:

1 Thermal expansion.
2 The thermoelectric effect.
3 Resistance change.
4 Radiative heat emission.

This list has excluded some other specialized devices which are expensive but give particularly high measurement accuracy in certain applications. Such devices include the quartz thermometer, based on resonant frequency change with temperature of a quartz crystal, the acoustic thermometer, based on changes in the velocity of sound with temperature, and fiber-optic devices. Details of these can be found elsewhere (e.g. Morris, 1993).

9.1.1 Thermal expansion methods

Thermal expansion methods make use of the fact that the dimensions of all substances, whether solids, liquids or gases, change with temperature. Instruments operating on this physical principle include the liquid-in-glass thermometer, the bimetallic thermometer and the pressure thermometer.

The *liquid-in-glass* thermometer is a well-known temperature measuring instrument which is used in a wide range of applications. The fluid used is usually either mercury or coloured alcohol, and this is contained within a bulb and capillary tube. As the temperature rises, the fluid expands along the capillary tube and the meniscus level is read against a calibrated scale etched on the tube. The process of estimating the position of the curved meniscus of the fluid against the scale introduces some error into the measurement process and a measurement inaccuracy lower than ±1% of full-scale reading is hard to achieve. Industrial versions of the liquid-in-glass thermometer are normally used to measure temperature in the range between −200 °C and +1000 °C, although instruments are available to special order which can measure temperatures up to 1500 °C.

The bimetallic principle is commonly used in thermostats. It is based on the fact that if two strips of different metals are bonded together, any temperature change will cause the strip to bend, as this is the only way in which the differing rates of change of length of each metal in the bonded strip can be accommodated. In the bimetallic thermostat, this is used as a switch in control applications. If the magnitude of bending is measured, the device becomes a *bimetallic thermometer*. The strip is often arranged in a spiral or helical configuration, as this gives a relatively large displacement of the free end for any given temperature change. The measurement sensitivity can be increased further by choosing the pair of materials carefully such that the degree of bending is maximized, with Invar (a nickel–steel alloy) and brass being commonly used. Bimetallic thermometers are used to measure temperatures between −75 °C and +1500 °C. The inaccuracy of the best instruments can be as low as ±0.5%, but devices of this quality are quite expensive. Many instrument applications do not require this degree of accuracy in temperature measurements, and in such cases much cheaper bimetallic thermometers with substantially inferior accuracy specifications are used.

The sensing element in a *pressure thermometer* consists of a bulb containing gas. If the gas were not constrained, temperature rises would cause its volume to increase. However, because it is constrained in the bulb and cannot expand, its pressure rises instead. As such, the pressure thermometer does not strictly belong to the thermal expansion class of instruments but is included because of the relationship between volume (V), pressure (P) and temperature (T) according to Boyle's gas law: $PV = KT$, where K is a constant. The change in pressure of the gas is measured by a suitable pressure transducer such as the Bourdon tube (see section 9.2). Pressure thermometers are used to measure temperatures in the range between −250 °C and +2000 °C. Their typical inaccuracy is ±0.5% of full-scale reading.

9.1.2 Thermoelectric effect instruments (thermocouples)

Thermoelectric effect instruments rely on the physical principle that, when any two different metals are connected together, an e.m.f., which is a function of the temperature, is generated at the junction between the metals. The e.m.f./temperature relationship is non-linear, but it approximates to a linear relationship over a reasonable temperature range for certain pairs of materials.

Wires of such pairs of materials are connected together at one end, and in this form are known as *thermocouples*. These form a very important class of device as they provide the most commonly used method of measuring temperatures in industry. Thermocouples are manufactured from various combinations of the base metals copper and iron, the base-metal alloys of alumel (Ni/Mn/Al/Si), chromel (Ni/Cr), constantan (Cu/Ni), nicrosil (Ni/Cr/Si) and nisil (Ni/Si/Mn), the noble metals platinum and tungsten, and the noble-metal alloys of platinum/rhodium and tungsten/rhenium. Only certain combinations of these are used as thermocouples and each standard combination is known by an internationally recognized type letter, for instance type K is chromel–alumel.

The materials used for thermocouples are produced so as to conform precisely with some defined composition specification. This ensures that the thermoelectric behaviour of the thermocouple is accurately known. The connection between the two wires is effected by welding, silver-soldering or in some cases just by twisting the wire ends together. Welding is the most common technique used generally, with silver-soldering being reserved for copper–constantan instruments.

The closed end of a thermocouple is known as its *hot junction*, and the open end where the output e.m.f. is measured is known as the *reference junction*. The e.m.f. measured is a function of the temperature difference between the hot and reference junctions. Therefore, the reference junction temperature must be known before the hot junction temperature can be calculated, and this is normally effected by holding it at 0 °C in a bath of melting ice. The temperature corresponding to the e.m.f. measured at the output of a thermocouple is normally calculated by interpolating between values in *thermocouple tables*. Thermocouple tables are produced by tabulating the e.m.f. and temperature values for standard thermocouple materials using the known physical e.m.f./temperature relationship for each material pair. Tables are contained in many texts (e.g. Morris, 1993), and always assume a zero reference junction temperature. In order to make a thermocouple conform to the precisely defined e.m.f./temperature characteristic described by standard tables, it is necessary that all metals used are refined to a high degree of pureness and all alloys are manufactured to an exact specification. This makes the materials used expensive, and consequently thermocouples are typically only a few centimetres long.

It is clearly impractical to connect a voltage measuring instrument to measure the thermocouple output in such close proximity to the environment whose temperature is being measured, and therefore extension leads up to several metres long are normally connected between the thermocouple and the measuring instrument. The materials used for extension leads have the same basic composition as the thermocouple but are manufactured to a lower specification which significantly reduces their cost. Such a solution is still prohibitively expensive in the case of noble metal thermocouples, and it is necessary in this case to search for base-metal extension leads which have a similar thermoelectric behaviour to the noble metal thermocouple. In this form, the extension leads are known as *compensating leads*. A typical example of this is the use of nickel/copper–copper extension leads connected to a platinum/rhodium–platinum thermocouple.

The five standard base-metal thermocouples are chromel–constantan (type E), iron–constantan (type J), chromel–alumel (type K), nicrosil–nisil (type N) and copper–constantan (type T). These are all relatively cheap to manufacture but they become inaccurate with age and have a short life.

Chromel–constantan devices give the highest measurement sensitivity of 80 μV/°C, with an inaccuracy of ±0.75% and a useful measuring range of −200 °C up to 900 °C. Unfortunately, while they can operate satisfactorily in oxidizing environments, their performance and life are seriously affected by reducing atmospheres.

Iron–constantan thermocouples have a sensitivity of 60 μV/°C and are the preferred type for general-purpose measurements in the temperature range of −150 °C to +1000 °C, where the typical measurement inaccuracy is ±1%. Their performance is little affected by either oxidizing or reducing atmospheres.

Copper–constantan devices have a similar measurement sensitivity of 60 μV/°C, but a typical inaccuracy of only ±0.5%, and find their main application in measuring sub-zero temperatures down to −200 °C . They are also used in both oxidizing and reducing atmospheres to measure temperatures up to 350 °C. Chromel–alumel thermocouples have a measurement sensitivity of only 45 μV/°C, although their characteristic is particularly linear over the temperature range between 700 °C and 1200 °C and this is therefore their main application. Like chromel–constantan devices, they are suitable for oxidizing atmospheres but not for reducing ones. Their typical measurement inaccuracy is ±0.75%.

Nicrosil–nisil thermocouples are a recent development which resulted from attempts to improve the performance and stability of chromel–alumel thermocouples. Their thermoelectric characteristic has a very similar shape to type K devices with equally good linearity over a large temperature measurement range and a measurement sensitivity of 40 μV/°C. The operating environment limitations are the same as for chromel–alumel devices but their long-term stability and life are at least three times better.

Noble-metal thermocouples are always expensive but enjoy high stability and long life in conditions of high temperature and oxidizing environments. They are chemically inert except in reducing atmospheres. Thermocouples made from platinum and a platinum/rhodium alloy have a low inaccuracy of ±0.2% and can measure temperatures up to 1500 °C, but their measurement sensitivity is only 10 μV/°C. Alternative devices made from tungsten and a tungsten/rhenium alloy have a better sensitivity of 20 μV/°C. and can measure temperatures up to 2300 °C.

Thermocouples, particularly base-metal types, are prone to contamination by various metals and protection is often necessary to minimize this. Whenever protection is required, it is achieved by enclosing the thermocouple in a sheath. Contamination alters the thermoelectric behaviour of thermocouples, such that their e.m.f./temperature characteristic varies from that published in standard tables. Contamination also makes them brittle and shortens their life.

Measurement sensitivity of thermocouples can be increased by combining several into a device known as a *thermopile*. This consists of several thermocouples connected together in series, such that all the reference junctions are at the same temperature and all the hot junctions are exposed to the temperature being measured. The effect of connecting n thermocouples together in series is to increase the measurement sensitivity by a factor of n. A typical thermopile manufactured by connecting together 25 chromel–constantan thermocouples gives a measurement resolution of 0.001 °C.

9.1.3 Varying resistance devices

Varying-resistance devices rely on the physical principle of the variation of resistance with temperature. The instruments are known as either resistance thermometers or thermistors according to whether the material used for their construction is a metal or a semiconductor material, and both are common measuring devices. The normal method of measuring resistance is to use a d.c. bridge. The excitation voltage of the bridge has to be chosen very carefully because, although a high value is desirable for achieving high measurement sensitivity, the self-heating effect of high currents in the temperature transducer creates an error by increasing the temperature of the device and so changing the resistance value.

Resistance thermometers, which are alternatively known as *resistance temperature devices (or RTDs)*, make use of the fact that the resistance of metals increases with temperature, albeit in a non-linear fashion. Fortunately, the resistance/temperature relationship is approximately linear over a limited temperature range for some metals, notably platinum, copper and nickel. Platinum is the most linear of these and it also has good chemical inertness: for these reasons the platinum resistance thermometer is preferred in most applications. Apart from having inferior linearity characteristics, both nickel and copper are very susceptible to oxidation and corrosion, which limits their accuracy and longevity. However, because platinum is very expensive compared with nickel and copper, nickel and copper resistance thermometers are sometimes used when cost is important. Another metal, tungsten, is also used in resistance thermometers in some circumstances, particularly for high-temperature measurements. The working range of each of these four types of resistance thermometer are shown below:

Platinum: −270 °C to +1000 °C (though use above 650 °C is uncommon)
Copper: −200 °C to +260 °C
Nickel: −200 °C to +430 °C
Tungsten: −270 °C to +1100 °C

In the case of non-corrosive and non-conducting environments, resistance thermometers are used without protection. In all other applications, they are protected inside a sheath.

Thermistors are manufactured from beads of semiconductor material prepared from oxides of the iron group of metals such as chromium, cobalt, iron, manganese, and nickel. All of these materials exhibit a large negative temperature coefficient, i.e. the resistance decreases as the temperature increases, which is fundamentally different from the positive temperature coefficient of the resistance thermometer. The non-linearity is high and it is not possible to make a linear approximation to the curve over even a small temperature range. The major advantages of thermistors are their relatively low cost and their small size. However, the size reduction also decreases its heat dissipation capability, and so makes the self-heating effect greater. In consequence, thermistors have to be operated at lower current levels than resistance thermometers, and so the measurement sensitivity afforded is less.

9.1.4 Radiation pyrometers (radiation thermometers)

All objects emit electromagnetic radiation as a function of their temperature above absolute zero and radiation pyrometers (also known as radiation thermometers) measure this radiation in order to calculate the temperature of the object. Different versions are capable of measuring temperatures between −20 °C and +1800 °C and give measurement inaccuracies as low as ±0.05%. Recently, portable, battery-powered, hand-held versions have become available which are particularly convenient to use. The important advantage that radiation pyrometers have over other types of temperature measuring instrument is that there is no contact with the hot body while its temperature is being measured. Thus, there is no disturbance of the measured system. Furthermore, there is no possibility of contamination, which is particularly important in food and many other process industries. They are especially suitable for measuring high temperatures that are beyond the capabilities of contact instruments such as thermocouples, resistance thermometers and thermistors. They are also capable of measuring moving bodies, for instance, the temperature of steel bars in a rolling mill.

Their use is not as straightforward as the discussion so far might have suggested, however, because the radiation from a body varies with the composition and surface condition of the body as well as with temperature. This dependence on surface condition is quantified by the *emissivity* of the body. Use is further complicated by absorption and scattering of the energy between the emitting body and the radiation detector. Therefore, all radiation pyrometers have to be carefully calibrated for each particular body whose temperature they are required to monitor.

Radiation pyrometers resemble a telescope in general structure, the main difference being that there is a radiation detector at the focal point of the lens system instead of an eyepiece. The radiation detector is either a thermal detector, which measures the temperature rise in a black body at the focal point of the optical system, or a photon detector. Thermal detectors respond equally to all wavelengths in the frequency spectrum whereas photon detectors respond selectively to a particular band within the full spectrum. Thermopiles, resistance thermometers and thermistors are all used as thermal detectors in different versions of these instruments. Photon detectors are usually of the photoconductive or photovoltaic type.

The minimum size of objects measurable by a radiation pyrometer is limited by the optical resolution, which is defined as the ratio of target size to distance — 1 : 300 is regarded as a good ratio, and this would allow temperature measurement of a 1 mm sized object at a range of 300 mm. With large distance/target size ratios, accurate aiming and focusing of the pyrometer at the target is essential. It is now common to find ‘through the lens’ viewing provided in pyrometers, using a principle similar to SLR (single lens reflex) camera technology, as focusing the instrument for visible light automatically focuses it for infrared light.

Thermography or *thermal imaging* uses the same principle as radiation pyrometers and involves scanning an infrared radiation detector across an object to measure the temperature distribution across it. Temperature measurement over the range from −20 °C up to +1500 °C is possible. Simpler versions of thermal imaging instruments consist of hand-held viewers which are pointed at the object of interest. Measurement resolution is high, with temperature differences as small as 0.1 °C being detectable. Such instruments are used in a wide variety of applications such as monitoring product flows through pipework, detecting insulation faults, and detecting hot spots in furnace linings, electrical transformers, machines, bearings, etc. Further information can be found in Morris (1993).

9.1.5 Intelligent temperature measuring instruments

Intelligent temperature transmitters have now been introduced into the catalogs of most instrument manufacturers, and they bring about the usual benefits associated with intelligent instruments. Such transmitters are separate boxes designed for use with temperature transducers which have either a d.c. voltage output in the mV range or an output in the form of a resistance change. They are therefore suitable for use in conjunction with thermocouples, thermopiles, resistance thermometers, thermistors and broad-band radiation pyrometers. All of the transmitters presently available have non-volatile memories, where all constants used in correcting output values for modifying inputs are stored, thus enabling the instrument to survive power failures without losing information.

Facilities in transmitters now available include adjustable damping, noise rejection, self-adjustment for zero and sensitivity drifts and expanded measurement range. These features can reduce the inaccuracy level to only ±0.05% of full scale.

The cost of intelligent temperature transmitters is significantly more than their non-intelligent counterparts, and justification purely on the grounds of their superior accuracy is hard to make. However, their expanded measurement range means immediate savings are made in terms of a reduction in the number of separate instruments needed to cover several measurement ranges. Their capability for self-diagnosis and self-adjustment also means that they require attention much less frequently, giving additional savings in maintenance costs.

9.2 Pressure measurement

Pressure measurement is a very common requirement in most industrial process control systems, and many different types of pressure-sensing and pressure-measurement systems are available. Pressure can be expressed as either absolute pressure, gauge pressure or differential pressure. The *absolute pressure* of a fluid defines the difference between the pressure of the fluid and the absolute zero of pressure, whereas *gauge pressure* describes the difference between the pressure of a fluid and atmospheric pressure. Absolute and gauge pressure are therefore related by the expression:

$$\text{absolute pressure} = \text{gauge pressure} + \text{atmospheric pressure}$$

The term *differential pressure* is used to describe the difference between two pressure values, such as the pressures at two different points within the same fluid, for example either side of a flow restrictor in a system measuring volume flow rate.

The range of pressures for which measurement is commonly required is the span from 1.013 to 7000 bar (1–6910 atmospheres). In the following discussion, therefore, devices used for measuring pressures in this common mid-range are described first, and this is followed by separate consideration of the techniques used for measuring pressure above and below this range.

9.2.1 **Measurement of mid-range pressures (1.013–7000 bar)**

The only instrument commonly used for measuring absolute mid-range pressures is the U-tube manometer. However, in the case of gauge and differential pressures, several instruments are used. Gauge pressure is generally measured in one of two ways, either by comparison with a known weight acting on a known area, or by deflection of elastic elements. Instruments belonging to the former class are the dead-weight gauge and the U-tube manometer, whereas the latter class consists of the diaphragm and Bourdon tube. Differential pressure can be measured either by a U-tube manometer or by an elastic-element type device (usually a diaphragm). Apart from the devices already mentioned, modern developments in electronics now permit the use of other principles in pressure measurement, such as in the resonant wire device.

Choice between the various types of instrument available for measuring mid-range pressures is usually strongly influenced by the intended application. U-tube manometers are very commonly used where visual indication of pressure levels is required, and dead-weight gauges, because of their superior accuracy, are used in calibration procedures of other pressure-measuring devices. Where compatibility with automatic control schemes is required, the choice of transducer is usually either a diaphragm type or a Bourdon tube, with the former now being predominant. Less frequently, bellows-type instruments are also used for this purpose, mainly in applications where their greater measurement sensitivity characteristics are required.

The *U-tube manometer* is a glass tube in the shape of a letter 'U' containing a fluid. When used for measuring absolute pressure, one end of the tube is sealed and evacuated, and the unknown pressure is applied to the open end of the tube. The absolute pressure is then measured in terms of the difference between the mercury levels in the two halves of the tube. When used for measuring gauge pressure, both ends of the tube are open, as shown in Figure 9.1, with the unknown pressure being applied at one end and the other end being open to the atmosphere. In the third mode of

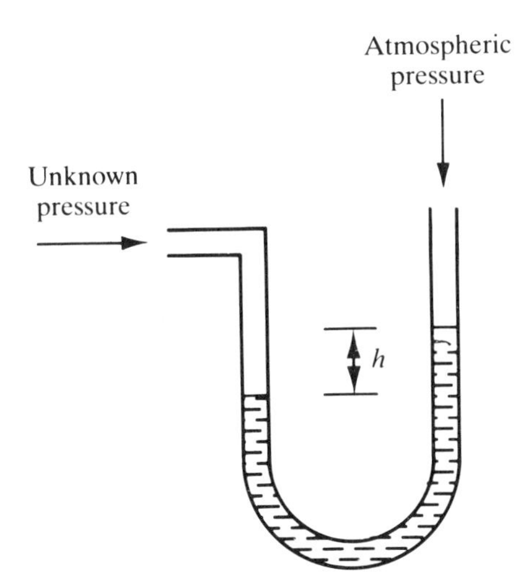

Figure 9.1 *Use of U-tube manometer to measure gauge pressure*

use, each open end of the tube is connected to different unknown pressures (p_1 and p_2), and the instrument thereby measures the differential pressure ($p_1 - p_2$).

U-tube manometers are typically used to measure gauge and differential pressures up to about 2 bar. The type of liquid used in the instrument depends on the pressure and characteristics of the fluid being measured. Water is a convenient and certainly cheap choice, but evaporates easily and is difficult to see. It is nevertheless used extensively, with the major obstacles to its use being overcome by using coloured water and regularly topping up the tube to counteract evaporation. The U-tube manometer, in one of its various forms, is an instrument commonly used in industry to give a visual measurement of pressure which can be acted upon by a human operator. It is not normally possible to transform the output of a U-tube manometer into an electrical signal, however, and so this instrument is not suitable for use as part of automatic control systems. The *well-type manometer (cistern manometer)* and the *inclined manometer (draft gauge)* are variations of the U-tube principle.

The *dead-weight gauge*, as shown in Figure 1.5, is a null-reading type of measuring instrument in which weights are added to the piston platform until an etched line on the piston is aligned with a fixed reference mark, at which time the downward force of the weights on top of the piston is balanced by the pressure exerted by the fluid beneath the piston. The fluid pressure is therefore calculated in terms of the weight added to the platform and the known area of the piston. The instrument offers the ability to measure pressures to a high degree of accuracy but is inconvenient to use. Its major application is as a reference instrument against which other pressure-measuring devices are calibrated.

The *diaphragm* is one of three types of elastic-element pressure transducer, and is shown schematically in Figure 9.2. Diaphragm-type instruments predominate in the

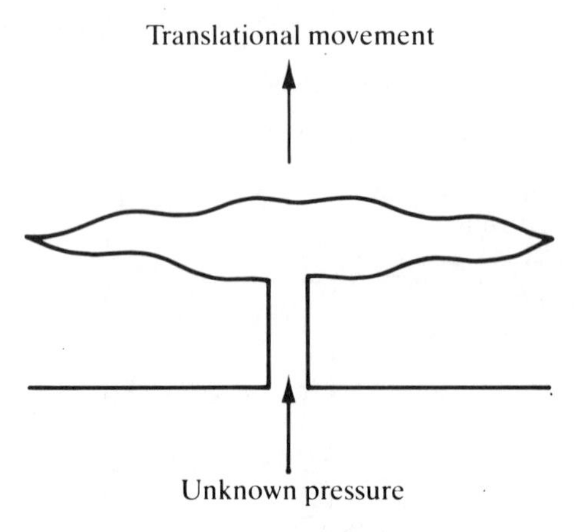

Figure 9.2 *Diaphragm*

measurement of pressures up to 10 bar. Applied pressure causes displacement of the diaphragm and this movement is measured by a displacement transducer. Both gauge pressure and differential pressure can be measured by different versions of diaphragm-type instruments. In the case of differential pressure, the two pressures are applied to either side of the diaphragm and the displacement of the diaphragm corresponds to the pressure difference. The typical magnitude of displacement in either version is 0.1 mm, which is well suited to a strain-gauge type of displacement-measuring transducer.

Four strain gauges are normally used in a bridge configuration, in which an excitation voltage is applied across two opposite points of the bridge. The output voltage measured across the other two points of the bridge is then a function of the resistance change due to the strain in the diaphragm. This arrangement automatically provides compensation for environmental temperature changes. Older pressure transducers of this type used metallic strain gauges bonded to a diaphragm typically made of stainless steel. Apart from manufacturing difficulties arising from the problem of bonding the gauges, metallic strain gauges in a d.c. bridge give a low-magnitude output voltage which has to be amplified by an expensive d.c. amplifier. The development of semiconductor, piezoresistive strain gauges provided a solution to the low-output problem, as they exhibit large resistance changes when strain is applied and give outputs which are up to one hundred times greater than those from metallic gauges. However, they do not overcome the difficulty of bonding them to a diaphragm and a new problem is introduced in respect of the highly non-linear characteristic of the strain/output relationship. The most recent development has been monolithic piezoresistive pressure transducers, which first appeared some ten years ago. These avoid the problem of bonding and they are now the most commonly used type of diaphragm pressure transducer. The monolithic cell consists of a diaphragm cut from a silicon sheet into which resistors are diffused during the manufacturing process. Such pressure transducers can be made to be very small and are often known as *micro-sensors*.

As an alternative to strain-gauge type displacement measurement, capacitive transducers are sometimes used. These also can be diffused into a silicon chip and fabricated as very small micro-sensors.

The *bellows* is a device which operates on a very similar principle to the diaphragm. It has a greater sensitivity than a diaphragm but is now rarely used because of its high manufacturing cost and proneness to failure.

The *Bourdon tube* is another type of elastic element pressure transducer and it is a very common industrial measuring instrument used for measuring the pressure of both gaseous and liquid fluids. It consists of a specially shaped piece of oval-section, flexible tube which is fixed at one end and free to move at the other. When pressure is applied at the open, fixed end of the tube, the oval cross-section becomes more circular. As the cross-section of the tube tends towards a circular shape, a deflection of the closed, free end of the tube is caused. This displacement is measured by some form of displacement transducer, which is commonly a potentiometer or LVDT (linear variable differential transformer), or less often a capacitive sensor (see section 9.5 on motion measurement for more details). In its most common form as a C-type bourdon tube, the tube shape resembles a letter 'C', as shown in Figure 9.3. *Spiral and helical Bourdon tubes* also exist. These give a greater deflection of the free end for a given applied pressure and therefore achieve greater measurement sensitivity and resolution. C-type tubes are available for measuring pressures up to 6000 bar, with measurement inaccuracy typically quoted as ±1% of full-scale deflection. Similar accuracy is available from helical and spiral types, but while the measurement resolution is higher, the maximum pressure measurable is only 700 bar.

The *resonant-wire device* is a relatively new instrument which has emanated from recent advances in the electronics field. Wire is stretched across a chamber containing fluid at unknown pressure subjected to a magnetic field. The wire resonates at its natural frequency according to its tension, which varies with pressure. Thus pressure is calculated by measuring the frequency of vibration of the wire. Such frequency measurement is normally carried out by electronics integrated into the cell. Measurement inaccuracy is very low, with ±0.2% full-scale reading being typical, and they are particularly insensitive to ambient condition changes.

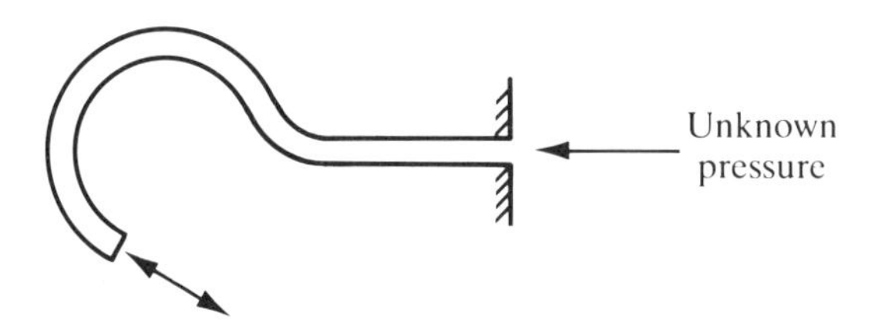

Figure 9.3 *C-type Bourdon tube*

In *fiber-optic pressure sensors*, also known as *microbend sensors*, the refractive index of the fiber and hence the intensity of light transmitted varies according to the mechanical deformation of the fibers caused by pressure. Microbend sensors are typically used to measure the small pressure changes generated in vortex-shedding flowmeters (see section 9.3.2).

9.2.2 **Low-pressure measurement (less than 1.013 bar)**

Adaptations of most of the common types of pressure transducer already described can be used for absolute pressure measurement in the vacuum range (less than atmospheric pressure). Special forms of Bourdon tubes measure pressures down to 10 mbar, manometers and bellows-type instruments measure pressures down to 0.1 mbar, and diaphragms can be designed to measure pressures down to 0.001 mbar. Other more specialized instruments are also available to measure vacuum pressures, and these generally give better accuracy than instruments which are primarily designed for measuring mid-range pressures. These special pressure gauges include the thermocouple gauge, the thermistor gauge, the Pirani gauge, the McLeod gauge and the ionization gauge.

Such specialized instruments are quite expensive, but they offer high measurement accuracy and sensitivity. At pressures between 0.001 mbar and 1 bar, thermocouple/thermistor gauges, McLeod gauges and Pirani gauges all find application in measurement situations which demand high accuracy and sensitivity. Below 0.001 mbar, these four instruments, together with the ionization gauge, form the only option. Thermocouple/thermistor gauges measure down to 10^{-4} mbar, Pirani gauges to 10^{-5}mbar and ionization gauges down to 10^{-13} mbar.

9.2.3 **High-pressure measurement (greater than 7000 bar)**

Measurement of pressures above 7000 bar is normally carried out electrically by monitoring the change of resistance of wires of special materials. Materials having resistance-pressure characteristics which are suitably linear and sensitive include gold–chromium alloys and manganin. A coil of such wire is enclosed in a sealed, kerosene-filled, flexible bellows. The unknown pressure is applied to one end of the bellows, which transmits the pressure to the coil. The magnitude of the applied pressure is then determined by measuring the coil resistance.

9.2.4 **Intelligent pressure transducers**

Adding microprocessor power to pressure transducers brings about substantial improvements in their characteristics. Measurement sensitivity improvement, extended measurement range, compensation for hysteresis and other non-linearities, and correction for ambient temperature and pressure changes are just some of the facilities offered by intelligent pressure transducers. Inaccuracies of ±0.1% or less can be achieved with piezoresistive-bridge silicon devices, for instance. In view of their much superior characteristics, it is perhaps surprising that intelligent pressure instruments only represent about 1% of the total number of pressure measuring devices sold at the present time. The significantly higher cost compared with non-intelligent devices has been cited as the reason for this, but this hardly seems adequate to explain such a very low level of market penetration.

9.3 **Flow measurement**

Flow measurement is extremely important in all the process industries. The manner in which the flow rate is quantified depends on whether the quantity flowing is a solid, liquid or gas. In the case of solids, it is appropriate to measure the mass flow rate, whereas in the case of liquids and gases, flow is usually measured in terms of the volume flow rate.

9.3.1 ***Mass flow rate***

Measurement of the mass flow rate of solids in the process industries is normally concerned with solids which are in the form of small particles produced by a crushing or grinding process. Such materials are usually transported by some form of conveyor, and this allows the mass flow rate to be calculated in terms of the mass of the material on a given length of conveyor multiplied by the speed of the conveyor.

9.3.2 ***Volume flow rate***

Volume flow rate is the appropriate way of quantifying the flow of all materials which are in a gaseous, liquid or semi-liquid slurry form where solid particles are suspended in a liquid host. Materials in these forms are carried in pipes, and the common classes of instrument used for measuring the volume flow rate can be summarized as follows:

1 Differential pressure meters.
2 Variable area meters.
3 Positive displacement meters.
4 Turbine flowmeters.
5 Electromagnetic flowmeters.
6 Vortex-shedding flowmeters.
7 Gate-type meters.
8 Ultrasonic flowmeters.

The number of relevant factors to be considered when specifying a flowmeter for a particular application is very large. These include the temperature and pressure of the fluid, its density, viscosity, chemical properties and abrasiveness, whether it contains particles and whether it is a liquid or a gas. The required performance factors of accuracy, measurement range, acceptable pressure drop, output signal characteristics, reliability and service life must also be assessed.

Differential pressure meters

Differential pressure meters involve the insertion of some device into a fluid-carrying pipe which causes an obstruction. When such a restriction is placed in a pipe, the velocity of the fluid through the restriction increases and the pressure decreases. The volume flow rate is then proportional to the square root of the pressure difference across the obstruction. The manner in which this pressure difference is measured is important. Measuring the two pressures with different instruments and calculating the difference between the two measurements is not satisfactory because of the large measurement error which can arise when the pressure difference is small, as explained in section 3.5.4. The normal procedure is therefore to use a diaphragm-based differential pressure transducer.

The *orifice plate* is by far the most commonly used obstruction device. It consists of a metal disk with a hole in it, as shown in Figure 9.4. The use of the orifice plate is so widespread because of its simplicity, cheapness and availability in a wide range of sizes. However, inaccuracy with this type of obstruction device is ±2% of full scale or greater, and it suffers from a number of problems. Firstly, the edges of the hole gradually get worn away, which affects the flow-rate/pressure-difference relationship. Secondly, there is a tendency for any particles in the flowing fluid to stick behind the hole and gradually build up and reduce its diameter. One further serious problem with the orifice plate is that it causes a permanent pressure loss in the fluid-carrying pipe. Where this pressure loss is unacceptable, other obstruction devices are used like the *venturi* shown in Figure 9.5, and similarly shaped devices such as the *flow nozzle* and the *Dall flow tube*. Because of their special design, such devices cause much less permanent pressure loss in the system and are much less prone to wear and particle build up, but their cost is very much greater than that of orifice plates.

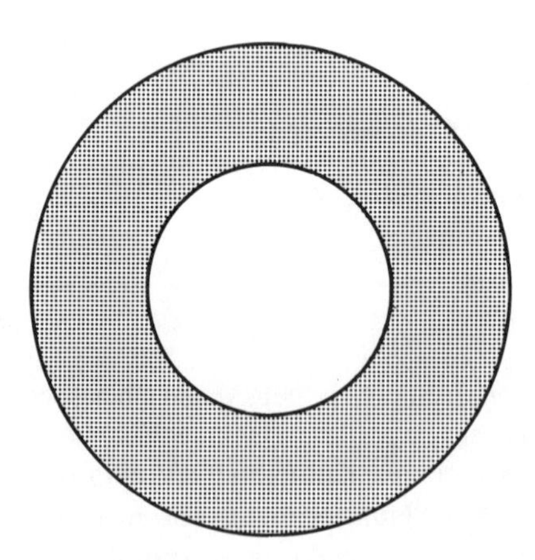

Figure 9.4 *The orifice plate*

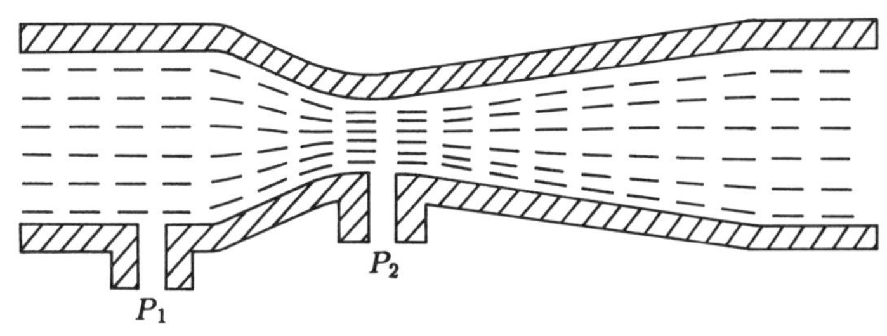

Figure 9.5 *The venturi*

Mention should also be made of the *Pitot tube* which is mainly used for making temporary measurements of flow, although it is also used in some instances for permanent flow monitoring. The Pitot tube is a very thin tube which obstructs only a small part of the flowing fluid, and so causes negligible pressure loss in the system. Because it only measures flow at a single point across the cross-section of the pipe, the measurement will only accurately represent the mean flow rate in the pipe if the flow of fluid is uniform across the pipe cross-section. The *Annubar* is a type of multi-port Pitot tube which does measure the average flow across the cross-section of the pipe by forming the mean value of several local flow measurements taken across the cross-section of the pipe.

Variable area flowmeters

In this class of flowmeter, the differential pressure across a variable aperture is used to adjust the area of the aperture. In its simplest form, shown in Figure 9.6, the instrument consists of a tapered glass tube containing a float which takes up a stable position where its submerged weight is balanced by the upthrust due to the differential pressure across it. The position of the float is a measure of the effective annular area of the flow passage and hence of the flow rate. Inaccuracy in the cheapest instruments is typically ±3%, but more expensive versions offer much lower measurement inaccuracies as small as ±0.2%. The normal measurement range is between 10% and 100% of the full-scale reading for any particular instrument.

This type of instrument normally only gives a visual indication of flow rate, and so is of no use in automatic control schemes. However, it is reliable and cheap and used extensively throughout industry. In fact, variable area meters account for 20% of all flowmeters sold.

Positive displacement flowmeters

This class of flowmeters, which accounts for nearly 10% of the total number of flowmeters used in industry, uses various mechanical arrangements to displace discrete volumes of fluid successively. They are all low friction, low maintenance and long life devices, although they do impose a small permanent pressure loss on the flowing fluid.

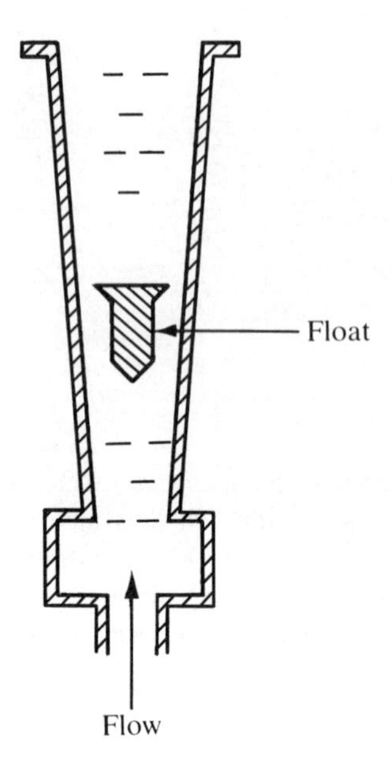

Figure 9.6 *The variable area flowmeter*

The *rotary piston meter* is a common type, and the principles of operation of this are shown in Figure 9.7. It consists of a slotted cylindrical piston moving inside a cylindrical working chamber which has an inlet port and an outlet port. The piston moves round the chamber such that its outer surface maintains contact with the inner surface of the chamber, and, as this happens, the piston slot slides up and down a fixed division plate in the chamber. At the start of each piston motion cycle, liquid is admitted to volume B from the inlet port. The fluid pressure causes the piston to start to rotate around the chamber, and, as this happens, liquid in volume C starts to flow out of the outlet port, and also liquid starts to flow from the inlet port into volume A. As the piston rotates further, volume B becomes shut off from the inlet port, while liquid continues to be admitted into A and pushed out of C. When the piston reaches the end-point of its motion cycle, the outlet port is opened to volume B, and the liquid which has been transported round inside the piston is expelled. After this, the piston pivots about the contact point between the top of its slot and the division plate, and volume A effectively becomes volume C ready for the start of the next motion cycle. A peg on top of the piston causes a reciprocating motion of the lever attached to it. This is made to operate a counter, and the flow rate is therefore determined from the count in unit time multiplied by the quantity (fixed) of liquid transferred between the inlet and outlet ports for each motion cycle.

Turbine meters

A turbine flowmeter consists of a multi-bladed wheel mounted in a cylinder along an

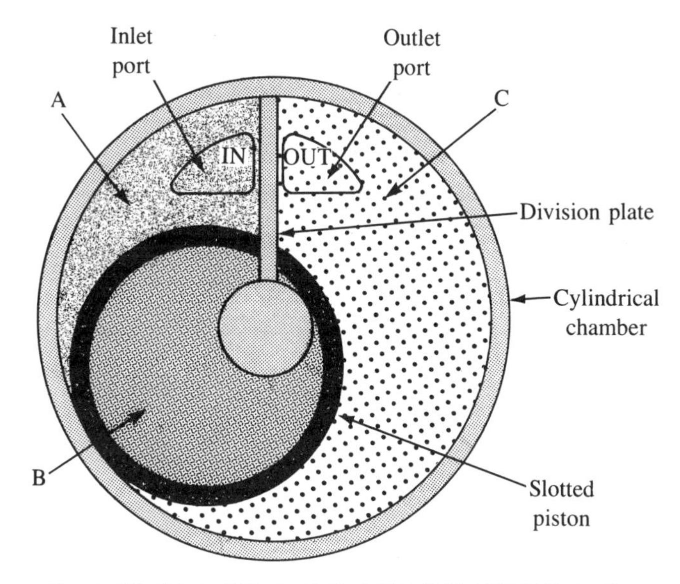

Figure 9.7 *Rotary piston form of positive displacement flowmeter*

axis parallel to the direction of fluid flow in the pipe. The flow of fluid past the wheel causes it to rotate at a rate which is proportional to the volume flow rate of the fluid. Provided that the turbine wheel is mounted in low-friction bearings, measurement inaccuracy can be as small as ±0.1%. However, turbine flowmeters are less rugged and less reliable than flow-restriction type instruments, and are badly affected by any particulate matter in the flowing fluid. Bearing wear is a particular problem and they also impose a permanent pressure loss on the measured system.

Turbine meters have a cost and market share which is similar to that of positive displacement meters, and compete for many applications, particularly in the oil industry. Turbine meters are smaller and lighter than positive displacement meters and are preferred for low-viscosity, high-flow measurements. However, positive-displacement meters are superior in conditions of high viscosity and low flow rate.

Electromagnetic flowmeters

Electromagnetic flowmeters are limited to measuring the volume flow rate of electrically conductive fluids. The measurement inaccuracy of around ±1.5% is quite acceptable, but the instrument is expensive both in terms of the initial purchase cost and also in running costs, mainly because of its electrical energy consumption. A further reason for high cost is the need for careful individual calibration of each instrument after manufacture, as there is considerable variation in the properties of the magnetic materials used. The instrument consists of a stainless steel cylindrical tube, fitted with an insulating liner, which carries the measured fluid. A magnetic field is created in the tube by placing mains-energized field coils either side of it, and the voltage induced in the fluid is measured by two electrodes inserted into opposite sides of the tube. The voltage measured is proportional to the flow rate. The internal diameter of electromagnetic flowmeters is normally the same as that of the rest of the flow-carrying pipework in the system. Therefore, there is no obstruction to the fluid flow and consequently no pressure loss associated with measurement. While the flowing fluid must be electrically conductive, the method is of use in many applications and is particularly useful for measuring the flow of slurries in which the liquid phase is electrically conductive. Corrosive fluids can be handled providing a suitable lining material is used. At the present time, electromagnetic flowmeters account for about 15% of the new flowmeters sold and this total is slowly growing.

Vortex-shedding flowmeters

The vortex-shedding flowmeter is a relatively new type of instrument which is rapidly gaining in popularity and is being used as an alternative to traditional differential pressure meters in more and more applications. The operating principle of the instrument is based on the natural phenomenon of vortex shedding, which arises when an unstreamlined obstacle, known as a bluff body, is placed in a fluid-carrying pipe, as indicated in Figure 9.8. When fluid flows past the obstacle, boundary layers of viscous, slow-moving fluid are formed along the outer surface. Because the obstacle is not streamlined, the flow cannot follow the contours of the body on the downstream side, and the separate layers become detached and roll into eddies or vortices in the low-pressure region behind the obstacle. The shedding frequency of these alternately shed vortices is proportional to the fluid velocity past the body. Various thermal, magnetic, ultrasonic and capacitive vortex-detection techniques are employed in different instruments. Such instruments have no moving parts, operate over a wide flow range, need only a relatively low power input and require little maintenance. They can measure both liquid and gas flows and a common inaccuracy figure quoted is ±1% of full-scale reading.

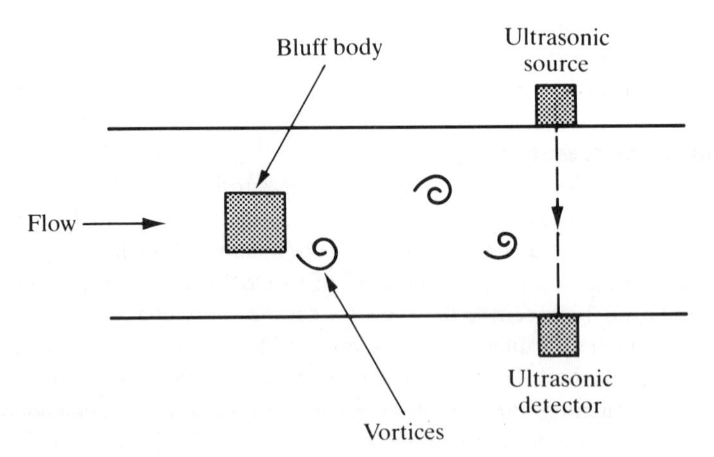

Figure 9.8 *The vortex-shedding flowmeter*

Gate-type meters

The gate meter was the earliest device in this class. It consists of a spring-loaded, hinged flap mounted at right angles to the direction of fluid flow in the fluid-carrying pipe. The flap is connected to a pointer outside the pipe. The fluid flow deflects the flap and pointer and the flow rate is indicated by a graduated scale behind the pointer. The major difficulty with such devices is in preventing leaks at the hinge point.

A variation on this principle is the *air-vane meter* which measures deflection of the flap by a potentiometer inside the pipe. This is commonly used to measure air flow within automotive fuel-injection systems.

Ultrasonic flowmeters

Ultrasonic flowmeters, like electromagnetic flowmeters, do not cause any pressure loss in the measured system. However, the ultrasonic technique of flow measurement is not restricted to conductive fluids (unlike electromagnetic flowmeters), and it is particularly useful for measuring the flow of corrosive fluids and slurries. A further advantage over electromagnetic flowmeters is that the instrument is one which clamps on externally to existing pipework rather than being inserted as an integral part of the flow line. As the procedure of breaking into a pipeline to insert a flowmeter can be as expensive as the cost of the flowmeter itself, the ultrasonic flowmeter has enormous cost advantages. Its clamp-on mode of operation has significant safety advantages in avoiding the possibility of personnel installing flowmeters coming into contact with hazardous fluids in the pipeline such as those which are poisonous, radioactive, flammable or explosive ones. Also, any contamination of the fluid being measured, for example food substances and drugs, is avoided.

The introduction of this type of flowmeter is a comparatively recent one and the present market share is therefore only about 1% of flowmeters sold. In view of their distinct advantages, however, this proportion is likely to increase over the next few years. Two different types of ultrasonic flowmeter exist which employ distinct technologies, one based on Doppler shift and the other on transit time. In the past, this has not always been readily understood, and has resulted in ultrasonic technology being rejected entirely when one of these two forms has been found to be unsatisfactory in a particular application. This is unfortunate, because the two technologies have distinct characteristics and areas of application, and many situations exist where even though one form is not suitable, the other is.

9.4 Level measurement

A wide variety of instruments is available for measuring the level of liquids and of solids which are in the form of powders or small particles. In most applications of such instruments, only an approximate indication of level is needed, and so the accuracy demands are not high.

9.4.1 **Dipsticks and float systems**

The *dipstick* is perhaps the simplest and cheapest device available for measuring liquid level. It consists of a metal bar, on which a scale is etched, which is fixed at a known position in the liquid-containing vessel. A level measurement is made by removing the instrument from the vessel and reading off how far up the scale the liquid has wetted. As a human operator is required to remove and read the dipstick, this method is limited to use in relatively small and shallow vessels. However, a more sophisticated form, known as an *optical dipstick*, allows a reading to be obtained without removing the dipstick from the vessel, and so is applicable to larger, deeper tanks. Measuring the position of a *float* on the surface of a liquid by means of a suitable transducer is another fairly simple and cheap method of liquid level measurement. In one version, a pivoted arm has a float at one end and a potentiometer at the other. Another version called the *float and tape gauge* (or *tank gauge*) has a tape attached to the float which passes round a pulley situated vertically above the float. The other end of the tape is attached to either a counterweight or a negative-rate counterspring. The amount of rotation of the pulley, measured by either a synchro or a potentiometer, is then proportional to the liquid level.

9.4.2 **Pressure measuring devices (hydrostatic systems)**

The hydrostatic pressure due to a liquid is directly proportional to its depth and hence to the level of its surface. Several instruments which use this principle for measuring liquid level are available and they are widely used in many industries, particularly in harsh chemical environments. In the case of open-topped vessels, or covered ones which are vented to the atmosphere, the level can be measured by inserting an appropriate pressure transducer at the bottom of the vessel. Alternatively, where liquid-containing vessels are totally sealed, the liquid level can be calculated by measuring the differential pressure between the top and bottom of the tank.

9.4.3 **Capacitive devices**

Capacitive devices are now widely used for measuring the level of liquids and of solids in powdered or granular form. They are suitable for use in extreme conditions, such as measuring high-temperature liquid metals, low-temperature liquid gases, corrosive liquids such as acids, and high-pressure processes. Two versions are used according to whether the measured substance is conducting or not. For non-conducting substances, two bare-metal capacitor plates in the form of concentric cylinders are immersed in the substance. The substance behaves as a dielectric between the plates according to the depth of the substance. In the case of conducting substances, the capacitor plates are encapsulated in an insulating material.

9.4.4 **Ultrasonic level gauge**

In this, energy from an ultrasonic source above the liquid is reflected back from the liquid surface into an ultrasonic energy detector. Measurement of the time of flight allows the liquid level to be inferred. In alternative versions, the ultrasonic source is placed at the bottom of the vessel containing the liquid, and the time of flight between emission, reflection off the liquid surface and detection back at the bottom of the vessel is measured. Ultrasonic techniques are especially useful in measuring the position of the interface between two immiscible liquids contained in the same vessel, or measuring the sludge or precipitate level at the bottom of a liquid-filled tank.

9.5 Motion measurement

Motion measurement encompasses measurements of displacement, velocity and acceleration in both translational and rotational forms.

9.5.1 *Translational displacement*

Apart from their use as a primary transducer measuring the motion of a body, translational displacement transducers are also widely used as a secondary component in measurement systems, where some other physical quantity such as pressure, force, acceleration or temperature is converted into a translational motion by the primary measurement transducer.

The *resistive potentiometer* is perhaps the best-known device in this class. The first versions of these had a wire-wound track but newer versions are now available with carbon film and plastic film tracks which give much better measurement resolution. The most common problem with potentiometers is dirt under the slider, which increases the resistance and thereby gives a false output voltage reading, or in the worst case causes a total loss of output. A typical inaccuracy figure quoted for translational motion resistive potentiometers is ±1% of full-scale reading. Manufacturers produce potentiometers to cover a large span of measurement ranges. At the bottom end of this span, instruments with a range of ±2 mm are available, while at the top end, instruments with a range of ±1 m are produced.

The *linear variable differential transformer*, which is commonly known by the abbreviation *LVDT*, consists of a transformer with a single primary winding and two secondary windings connected in the series-opposing manner shown in Figure 9.9. The object whose translational displacement is to be measured is physically attached to the central iron core of the transformer, so that all motions of the body are transferred to the core. Because of the series opposition mode of connection of the secondary windings, the output voltage is zero when the core is in the central position, but non-zero when the core is displaced. The only moving part in an LVDT is the core. As this is only moving in the air gap between the windings, there is no friction or wear during operation. For this reason, the instrument is a very popular one for measuring linear displacements and has a quoted life expectancy of 200 years. The typical inaccuracy is ±0.5% of full-scale reading and the measurement resolution is almost infinite.

Variable-inductance transducers, which are also sometimes referred to as *variable-reluctance* or *variable-permeance* transducers, exist in two separate forms.

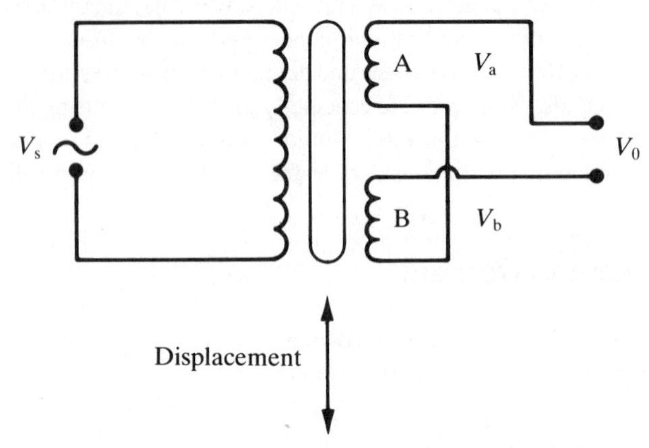

Figure 9.9 *The linear variable differential transformer (LVDT)*

Both of these rely for their operation on the variation of the mutual inductance between two magnetically coupled parts, one of which is attached to the body whose displacement is being measured. Their typical inaccuracy is ±0.5% of full scale. This is better than that of the LVDT but, in spite of this, their use is very much less common than that of the LVDT.

Variable capacitance transducers work on similar principles and have inaccuracies as small as ±0.01% and measurement resolutions as small as 1 μm. Individual devices can be selected from manufacturers' ranges which measure displacements as small as 10^{-11} m or as large as 1 m.

The *strain gauge* is a device which experiences a change in resistance when it is stretched or strained. It consists physically of a length of resistance wire formed into a zig-zag pattern and mounted on a flexible backing sheet, as shown in Figure 9.10. The wire is nominally of circular cross-section. As strain is applied to the gauge, the shape of the cross-section of the resistance wire distorts, changing the cross-sectional area. As the resistance of the wire per unit length is inversely proportional to the cross-sectional area, there is a consequential change in resistance. In use, strain gauges are bonded to the object whose displacement is to be measured. The strain gauge has a typical range of measurement of 0–50 μm and a very common application of strain gauges is in measuring the very small displacements which occur in the diaphragms of pressure-measuring instruments. Measurement inaccuracies as low as ±0.15% of full-scale reading are achievable.

9.5.2 **Translational velocity**

Translational velocity cannot be measured directly and is usually calculated either by differentiation of translational displacement measurements or by integration of an acceleration measurement. The latter is preferable because differentiation amplifies noise and so gives a poor-quality velocity measurement. Alternatively, translational

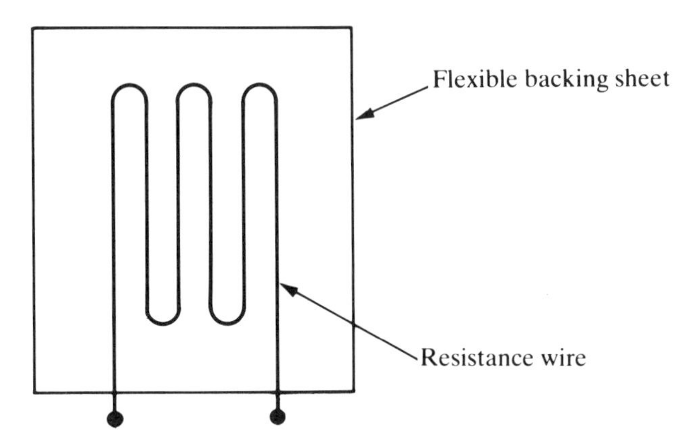

Figure 9.10 *The strain gauge*

motion can be converted to rotational velocity by suitable gearing and measured in that form.

9.5.3 **Rotational displacement**

The *circular potentiometer* is the cheapest device available and works on exactly the same principles as the translational motion potentiometer except that the track is bent round into a circular shape. In this form, the maximum measurement range is 360 degrees. A greater measurement range and resolution is obtained in the *helical potentiometer*, in which the track is formed into a helix shape. Inaccuracy varies from ±1% of full scale for circular potentiometers down to ±0.002% of full scale for the best helical potentiometers. As with linear track potentiometers, all rotational potentiometers can give performance problems because of dirt on the track.

The *rotational differential transformer* is a special form of differential transformer which measures rotational rather than translational motion. Like its linear equivalent, the instrument suffers no wear in operation and therefore has a very long life with almost no maintenance requirements. However, the lowest inaccuracy achievable is ±1% and this only applies for limited excursions of the core of ±40° away from the null position. For angular displacements of ±60°, the inaccuracy increases to ±3%, and the instrument is unsuitable for measuring displacements greater than this.

Incremental shaft encoders are one of a class of devices which give an output in digital form. They measure the instantaneous angular position of a shaft relative to some arbitrary datum point, but are unable to give any indication about the absolute position of a shaft. The principle of operation is to generate pulses as the shaft whose displacement is being measured rotates. These pulses are counted and the total angular rotation inferred from the pulse count. The pulses are generated either by optical or by magnetic means and are detected by suitable sensors. Of the two, the optical system is considerably cheaper and therefore much more common. The optical encoder is essentially a disk with a series of windows etched on it. A light source on one side of the disk produces pulses of light as the windows move past it. Directional information is provided by an inner set of windows which produce pulses which are out of phase with the main pulses, lagging or leading depending on whether the rotation is clockwise or anticlockwise.The maximum measurement resolution obtainable is limited by the number of windows which can be machined onto a disk. The maximum number of windows per track for a 150 mm diameter disk is 5000, which gives a basic angular measurement resolution of 1 in 5000.

Coded disk shaft encoders also give a digital output. However, unlike the incremental shaft encoder which gives an output in the form of pulses that have to be counted, the digital shaft encoder has an output in the form of a binary number of several digits which provides an absolute measurement of shaft position. Three different forms of digital shaft encoder exist using respectively optical, electrical and magnetic energy systems. The optical digital shaft encoder is the cheapest form available. The instrument is similar in physical appearance to the incremental shaft encoder, having a pair of disks, one movable and one fixed, with a light source on one side and light detectors on the other side. The fixed disk has a single window as in the incremental shaft encoder, and the principal way in which the device differs from the incremental shaft encoder is in the design of the windows on the movable disk. These are cut in four or more tracks instead of two and are arranged in sectors as well as tracks. An energy detector is aligned with each track, which gives an output of '1' when energy is detected and an output of '0' otherwise. The measurement resolution obtainable depends on the number of tracks used. For a four-track version, the resolution is one in sixteen, with progressively higher measurement resolution being attained as the number of tracks is increased. These binary outputs from the detectors are combined together to give a binary number of several digits. The number of digits corresponds to the number of tracks on the disk. The pattern of windows in each sector is cut such that, as that particular sector passes across the window in the fixed disk, the four energy detector outputs combine to give a unique binary number. It is possible to manufacture optical digital shaft encoders with up to twenty-one tracks. Such devices have a measurement resolution of 1 part in 10^6 (about one second of arc). However, such devices are very expensive.

Many other devices are also sometimes used for measuring rotational displacements. The *resolver* (also known as the *synchro-resolver*) and the *synchro* are electromechanical, motor-like devices, which have no contacting moving surfaces and so are frictionless and reliable in operation, with the best devices giving measurement resolutions of 0.1%. *Gyroscopes* measure both absolute angular displacement and absolute angular velocity.

9.5.4 **Rotational velocity**

If an angular acceleration transducer is available, such as a gyro-accelerometer, its output can be integrated to give an angular velocity measurement. Theoretically, differentiation of an angular displacement measurement could also be performed, but the resulting signal is usually too noisy to be of use because the differentiation process amplifies noise.

The other main instruments used are d.c. and a.c. tachometric generators (tachos). Alternatively, pulses can be generated by photoelectric techniques and counted.

9.5.5 *Measurement of acceleration*

Most forms of accelerometer consist of a mass suspended by a spring and damper inside a housing. If the accelerometer housing is rigidly fastened to an object, any acceleration of the object results in a force which causes a displacement of the mass. Various transducers are used to measure the mass displacement in different types of accelerometer, such as the potentiometer, strain gauge, LVDT, variable inductance transducer and piezo-electric transducer. Recently, very small *microsensors* have also become available for measuring acceleration. These consist of a small mass subject to acceleration which is mounted on a thin silicon membrane. Displacements are measured either by piezo-resistors deposited on the membrane or by etching a variable capacitor plate into the membrane.

One important characteristic of accelerometers is their sensitivity to accelerations at right angles to the sensing axis, which is the direction along which the instrument is designed to measure acceleration. This is defined as the *cross-sensitivity* and is specified in terms of the output, expressed as a percentage of the full-scale output, when an acceleration of some specified magnitude (e.g. 30 g) is applied at 90° to the sensing axis.

Most accelerometers measure translational acceleration, but the gyro accelerometer is used to measure rotational motion.

9.6 Vibration and shock measurement

9.6.1 *Vibration*

Vibrations normally consist of linear harmonic motion which can be expressed mathematically as:

$$X = X_o \sin(\omega t) \qquad (9.1)$$

where X is the displacement from the equilibrium position at any time t, X_o is the peak displacement from the equilibrium position, and ω is the angular frequency of the oscillations.

By differentiating equation (9.1) with respect to time, an expression for the velocity of the vibrating body at any time t is obtained and if it is differentiated again, an expression for the acceleration of the body is obtained. Thus, the intensity of vibration can be measured in terms of either displacement, velocity or acceleration.

Acceleration is clearly the best parameter to measure at high frequencies. However, because displacements are large at low frequencies according to equation (9.1), it would seem that measuring either displacement or velocity would be best at low frequencies. However, this is rarely done in practice because there are considerable practical difficulties in mounting and calibrating displacement and velocity transducers on a vibrating body. The frequency response of accelerometers used in vibration measurement is particularly important in view of the inherently high-frequency characteristics of the measurement environment. The most common type of transducer used is therefore the piezo-accelerometer, which uses a piezoelectric crystal and has a bandwidth which can be as high as 7 kHz.

We have thus explained why accelerometers are usually the primary component in vibration measurement. However, as well as an accelerometer, a vibration measurement system requires other elements to translate the accelerometer output into a recorded signal. The three other necessary elements are a signal-conditioning element, a signal analyzer and a signal recorder. The signal-conditioning element amplifies the relatively weak output signal from the accelerometer and also transforms the high output impedance of the accelerometer to a lower impedance value. The signal analyzer then converts the signal into the required output form, which may be either the peak value, r.m.s. value or average absolute value. The final element of the measurement system is then the signal recorder.

9.6.2 ***Shock***

Shock describes a type of motion where a moving body is brought suddenly to rest, often because of a collision. Shocks characteristically involve large-magnitude decelerations (e.g. 500 g) which last for a very short time (e.g. 5 ms). An instrument having a very high-frequency response is required for shock measurement, and for this reason, piezo-crystal-based accelerometers are commonly used.

Again, other elements for analyzing and recording the output signal are required, as for vibration measurement. A storage oscilloscope is a suitable instrument for recording the output signal, as this allows the time duration as well as the acceleration levels in the shock to be measured. Alternatively, if a permanent record is required, the screen of a standard oscilloscope can be photographed. A further option is to record the output on magnetic tape, which facilitates computerized signal analysis.

9.7 **Mass, force and torque measurement**

9.7.1 ***Mass***

One way in which the mass of a body can be measured is to compare the gravitational force on the body with the gravitational force on another body of known mass, using a beam balance, weigh beam, pendulum scale or electromagnetic balance, in a procedure known as 'weighing'. Alternative mass-measurement techniques are to use either a spring balance or a load cell. Of these, the electronic load cell has definite advantages and is the preferred instrument nowadays in more and more industrial applications.

In the *beam balance*, standard masses are added to a pan on one side of a pivoted beam until the magnitude of the gravity force on them balances the magnitude of the gravitational force on the unknown mass acting at the other end of the beam. This equilibrium position is indicated by a pointer which moves against a calibrated scale. The smallest measurement inaccuracy figure attainable is ±0.002%. One serious disadvantage of this type of instrument is its lack of ruggedness. Continuous use and the inevitable shock loading that will occur from time to time both cause damage to the knife edges, leading to problems in measurement accuracy and resolution. The *weigh beam* operates on similar principles to the beam balance but is much more rugged. The *pendulum scale* is also rugged and can be converted into a form which gives an electrical output. The *electromagnetic balance* is a further mass-balance instrument which has particular advantages in terms of its small size, insensitivity to environmental changes (modifying inputs) and its electrical form of output.

The *spring balance* is a well-known instrument for measuring mass. The mass is hung on the end of a spring and the deflection of the spring due to the downwards gravitational force on the mass is measured against a scale. Because the characteristics of the spring are very susceptible to environmental changes, measurement accuracy is usually relatively poor.

Electronic load cells, also known as *electronic balances*, have significant advantages over most other forms of mass-measuring instrument and so are the preferred type for most industrial applications. These advantages include relatively low cost, wide measurement range, tolerance of dusty and corrosive environments, remote measurement capability, tolerance of shock loading and ease of installation. The electronic load cell uses the physical principle that a force applied to an elastic element produces a measurable deflection. The elastic elements used are specially shaped and designed. The best accuracy in elastic force transducers is obtained by those instruments which use strain gauges to measure displacements of the elastic element, with an inaccuracy figure less than ±0.05% of full scale reading being obtainable.

Alternative forms of load cell also exist which work on either *pneumatic* or *hydraulic* principles, and convert mass measurement into a pressure measurement problem. Hydraulic load cells can attain the same level of accuracy as electronic load cells, but the inaccuracy of pneumatic load cells is usually higher, with ±0.5% of full scale being typical.

9.7.2 **Force**

If a force of magnitude F is applied to a body of mass m, the body will accelerate at a rate A according to the equation: $F = mA$. One way of measuring an unknown force is therefore to measure the acceleration when it is applied to a body of known mass. An alternative technique is to measure the variation in the resonant frequency of a vibrating wire as it is tensioned by an applied force.

The technique of applying a force to a known mass and measuring the acceleration produced is of very limited practical value because, in most cases, forces are not free entities but are part of a system from which they cannot be decoupled and in which they are acting on some body which is not free to accelerate. However, the technique can be of use in measuring some transient forces, and also for calibrating the forces produced by thrust motors in space vehicles. The *vibrating wire sensor* is more generally applicable. In this, the force is applied to a wire which is kept vibrating at its resonant frequency by a variable-frequency oscillator. As the wire tension changes, its resonant frequency also changes. Thus the force can be calculated by measuring the oscillator output frequency.

9.7.3 **Torque**

The three traditional methods of measuring torque consist of (a) measuring the reaction force in cradled shaft bearings, (b) the 'Prony brake' method and (c) measuring the strain produced in a rotating body due to an applied torque. These are all well described in many texts (e.g. Morris, 1993).

More recently, a new optical technique has become available which has particular advantages in terms of its relatively low cost and small physical size. The principles of the system are shown in Figure 9.11. Two black-and-white striped wheels are mounted at either end of the rotating shaft and are in alignment when no torque is applied to the shaft. Light from a laser diode light source is directed by a pair of fiber-optic cables onto the wheels. The rotation of the wheels causes pulses of reflected light and these are transmitted back to a receiver by a second pair of fiber-optic cables. Under zero torque conditions, the two pulse-trains of reflected light are in phase with each other. If torque is now applied to the shaft, the reflected light is modulated. Measurement by the

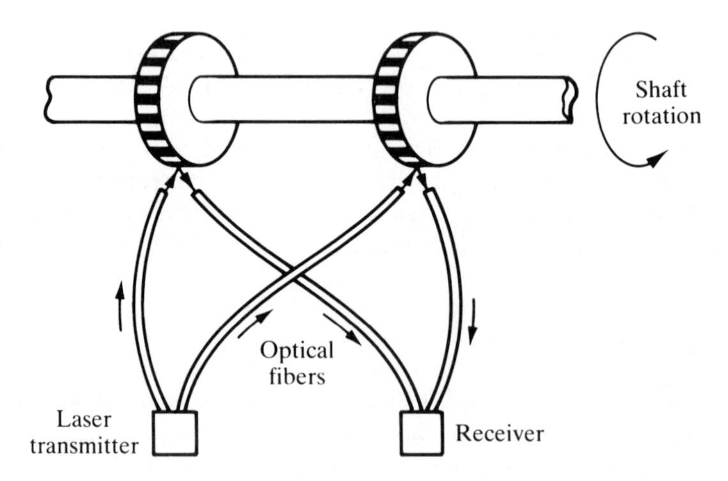

Figure 9.11 *Optical torque measurement*

receiver of the phase difference between the reflected pulse trains therefore allows the magnitude of torque in the shaft to be calculated.

References and further reading

Morris, A.S., *Principles of Measurement and Instrumentation*, 1993. Prentice Hall, Hemel Hempstead.

APPENDIX 1

Error function tables (F(z) where $z = (x-\mu)/\sigma$ (area under a normalized Gaussian curve)

	0.00	0.01	0.02	0.03	0.04	0.05	0.06	0.07	0.08	0.09
0.0	0.5000	0.5040	0.5080	0.5120	0.5160	0.5199	0.5239	0.5279	0.5319	0.5359
0.1	0.5398	0.5438	0.5478	0.5517	0.5557	0.5596	0.5636	0.5675	0.5714	0.5753
0.2	0.5793	0.5832	0.5871	0.5910	0.5948	0.5987	0.6026	0.6064	0.6103	0.6141
0.3	0.6179	0.6217	0.6255	0.6293	0.6331	0.6368	0.6406	0.6443	0.6480	0.6517
0.4	0.6554	0.6591	0.6628	0.6664	0.6700	0.6736	0.6772	0.6808	0.6844	0.6879
0.5	0.6915	0.6950	0.6985	0.7019	0.7054	0.7088	0.7123	0.7157	0.7190	0.7224
0.6	0.7257	0.7291	0.7324	0.7357	0.7389	0.7422	0.7454	0.7486	0.7517	0.7549
0.7	0.7580	0.7611	0.7642	0.7673	0.7703	0.7734	0.7764	0.7793	0.7823	0.7852
0.8	0.7881	0.7910	0.7939	0.7967	0.7995	0.8023	0.8051	0.8078	0.8106	0.8133
0.9	0.8159	0.8186	0.8212	0.8238	0.8264	0.8289	0.8315	0.8340	0.8365	0.8389
1.0	0.8413	0.8438	0.8461	0.8485	0.8508	0.8531	0.8554	0.8577	0.8599	0.8621
1.1	0.8643	0.8665	0.8686	0.8708	0.8729	0.8749	0.8770	0.8790	0.8810	0.8830
1.2	0.8849	0.8869	0.8888	0.8906	0.8925	0.8943	0.8962	0.8980	0.8997	0.9015
1.3	0.9032	0.9049	0.9066	0.9082	0.9099	0.9115	0.9131	0.9147	0.9162	0.9177
1.4	0.9192	0.9207	0.9222	0.9236	0.9251	0.9265	0.9279	0.9292	0.9306	0.9319
1.5	0.9332	0.9345	0.9357	0.9370	0.9382	0.9394	0.9406	0.9418	0.9429	0.9441
1.6	0.9452	0.9463	0.9474	0.9484	0.9495	0.9505	0.9515	0.9525	0.9535	0.9545
1.7	0.9554	0.9564	0.9573	0.9582	0.9591	0.9599	0.9608	0.9616	0.9625	0.9633
1.8	0.9641	0.9648	0.9656	0.9664	0.9671	0.9678	0.9686	0.9693	0.9699	0.9706
1.9	0.9713	0.9719	0.9726	0.9732	0.9738	0.9744	0.9750	0.9756	0.9761	0.9767
2.0	0.9772	0.9778	0.9783	0.9788	0.9793	0.9798	0.9803	0.9808	0.9812	0.9817
2.1	0.9821	0.9826	0.9830	0.9834	0.9838	0.9842	0.9846	0.9850	0.9854	0.9857
2.2	0.9861	0.9864	0.9868	0.9871	0.9875	0.9878	0.9881	0.9884	0.9887	0.9890
2.3	0.9893	0.9896	0.9898	0.9901	0.9904	0.9906	0.9909	0.9911	0.9913	0.9916
2.4	0.9918	0.9920	0.9922	0.9924	0.9927	0.9929	0.9930	0.9932	0.9934	0.9936
2.5	0.9938	0.9940	0.9941	0.9943	0.9945	0.9946	0.9948	0.9949	0.9951	0.9952
2.6	0.9953	0.9955	0.9956	0.9957	0.9959	0.9960	0.9961	0.9962	0.9963	0.9964
2.7	0.9965	0.9966	0.9967	0.9968	0.9969	0.9970	0.9971	0.9972	0.9973	0.9974
2.8	0.9974	0.9975	0.9976	0.9977	0.9977	0.9978	0.9979	0.9979	0.9980	0.9981
2.9	0.9981	0.9982	0.9982	0.9983	0.9984	0.9984	0.9985	0.9985	0.9986	0.9986
3.0	0.9986	0.9987	0.9987	0.9988	0.9988	0.9989	0.9989	0.9989	0.9990	0.9990
3.1	0.9990	0.9991	0.9991	0.9991	0.9992	0.9992	0.9992	0.9992	0.9993	0.9993
3.2	0.9993	0.9993	0.9994	0.9994	0.9994	0.9994	0.9994	0.9995	0.9995	0.9995
3.3	0.9995	0.9995	0.9995	0.9996	0.9996	0.9996	0.9996	0.9996	0.9996	0.9996
3.4	0.9997	0.9997	0.9997	0.9997	0.9997	0.9997	0.9997	0.9997	0.9997	0.9998
3.5	0.9998	0.9998	0.9998	0.9998	0.9998	0.9998	0.9998	0.9998	0.9998	0.9998
3.6	0.9998	0.9998	0.9998	0.9999	0.9999	0.9999	0.9999	0.9999	0.9999	0.9999

APPENDIX 2

Solutions to selected self-assessment questions

Chapter 2

2.3 0.0175 mV/°C

2.5 (a) 2.62

(b) 2.94; 0.32

2.6 (a) 20 μm/kg ; 22 μm/kg

(b) 200 μm ; 2 μm/kg

(c) 14.3 μm/°C ; 0.143 μm$(°C)^{-1}(kg)^{-1}$

2.7

(a)

Time (s)	Depth (m)	Temp. reading (°C)	Temp. error (°C)
0	0	20.0	0.0
100	50	19.72	0.22
200	100	19.25	0.25
300	150	18.75	0.25
400	200	18.25	0.25
500	250	17.75	0.25

(b) 10.25 °C

Chapter 3

3.3 Error = 6.25% ($E_m/E_0 = 0.9375$)

3.7 Mean 31.1; median 30.5; standard deviation 3.0

3.8 Mean 1.537; standard deviation 0.021;
accuracy of mean value = ±0.007,
i.e. mean value = 1.537±0.007;
accuracy improved by factor of 10

3.9 86.6%

3.10 97.7%
3.11 ±0.7%
3.12 ±4.7%
3.13 ±3%
3.14 46.7 Ω ±1.3%
3.15 4.5%
3.16 (a) 0.31 m^3/min; (b) ± 5.6%

Chapter 6

6.1 (a)

$$V_o = V_s\left(\frac{R_1}{R_1 + R_4} - \frac{R_2}{R_2 + R_3}\right)$$

(b) 81.9 mV
6.2 378 mV
6.3 (a)

$$V_o = V_i\left(\frac{R_u}{R_u + R_3} - \frac{R_1}{R_1 + R_2}\right)$$

(b) 0.82 mV/°C
(c) indicated temperature 101.9 °C ; error 1.9 °C
6.4 24 V; 1.2 W
6.6 (a) 69.6 Ω, 930.4Ω; (b) 110.3Ω
6.7 (a) $R_u = R_2R_3/R_1; L_u = R_2R_3C$
(b) 1.57Ω; 100 mH
(c) 20
6.8 2.538 Vrms
6.9 50 μF
6.10 (a) at balance

$$\frac{R_1 + j\omega L}{R_3} = \frac{R_2}{R_4 - j/\omega C}$$

by taking real and imaginary parts and manipulating:

$$L = R_1/\omega^2 R_4 C \qquad \text{(A)}$$

$$R_1 = \frac{R_2R_3}{R_4(1 + 1/\omega^2 R_4{}^2 C^2)}$$

hence, at balance,

$$L = \frac{R_2 R_3 C}{1 + \omega^2 {R_4}^2 C^2} \qquad \text{(B)}$$

(b) $Q = \omega L/R_1 = 1/\omega R_4 C$ using equation (A) above.
For large Q, $\omega^2 {R_4}^2 C^2 \ll 1$, and equation (B) above becomes:

$$L = R_2 R_3 C$$

This is independent of frequency because there is no ω term in the expression.
(c) 20 mH

Chapter 7

7.2 (a) moving-coil (mean) 0.685 A
(b) moving-iron (r.m.s.) 6.455 A
7.3 (a) 3.18 A
(b) 7.07 A ; 6.37 A
7.4 (i) mean 6.27 V (ii) r.m.s. 6.46 V
7.5 (a) 1.73 V; (b) 1.67 V; (c) 2.12 V
7.6 (b) (i) dynamometer (r.m.s.) 0.583 A ; moving coil (mean) 0.267 A; (ii) 68W

Chapter 8

8.1 (b) $\omega_n = \sqrt{K_S/J}$; $\beta = {K_I}^2/[2R\sqrt{JK_S}]$; sensitivity = $K_I/(K_S R)$
(d) 0.7
(e) typical bandwidth 100 Hz; maximum frequency 30 Hz
8.2 a = 12.410; b= 40.438
8.3 9.8 Ω
8.4 a = 1.12; b = 2.00
8.5 (a) $C = 5.77 \times 10^{-7}$; T_o = 11,027
(b) T = 428 K

Index